ECONOMIC ADJUSTMENT IN OIL-BASED ECONOMIES

To my parents

Economic Adjustment in Oil-based Economies

AHMAD JAZAYERI

HD
9576
.I62
J39
1988
West

ASU WEST LIBRARY

Avebury

Aldershot · Brookfield USA · Hong Kong · Singapore · Sydney

© Ahmad Jazayeri 1988

All rights reserved. No part of this publication may be reproduced, stored in a retrieval system, or transmitted in any form or by any means, electronic, mechanical, photocopying, recording, or otherwise without the prior permission of the Gower Publishing Company Limited.

Published by
Avebury
Gower Publishing Company Limited
Gower House
Croft Road
Aldershot
Hants GU11 3HR
England

Gower Publishing Company
Old Post Road
Brookfield
Vermont 05036
USA

British Library Cataloguing in Publication Data
Jazayeri, Ahmad, 1957–
  Economic adjustment in oil-based economics.
  1. Iran. Economic development. Role of Petroleum industries, 1970-1982
  2. Nigeria. Economic development. Role of petroleum industries, 1970-1982
  I. Title
  338.2'7282'0955

Library of Congress Cataloging–in–Publication Data
Jazayeri, Ahmad, *1957–*
  Economic adjustment in oil-based economies / Ahmad Jazayeri.
    p. cm.
  Bibliography: p.
  Includes index
  1. Petroleum industry and trade–Iran. 2. Iran–Manufactures.
3. Agriculture–Economic aspects–Iran. 4. Petroleum industry and trade–Nigeria.
5. Agriculture–Economic aspects–Nigeria.
I. Title.
HD9576.I62J39 1988
330.955'054--dc19   88-14057
ISBN 0 566 05682 8   CIP

Printed and bound in Great Britain by
Athenaeum Press Limited, Newcastle upon Tyne

Contents

Acknowledgements vii

Introduction ix

PART I THEORY

1 The Theoretical Framework 1

Non-traded goods	2
The core adjustment model	4
Factor incomes adjustment	11
Money, relative prices, and payments adjustment	12
The role of capital inputs and the cost of production	16
The role of government	19

PART II THE CASE OF IRAN

2 Adjustment Policy, Development Strategy and Macroeconomic Impact on Iran in the 1970s 35

Oil revenues and changes in aggregate expenditure 36

The development strategy and the composition of expenditure	39
The non-oil Gross Domestic Product and its components	48
Changes in the sectoral distribution of GDP	53
Conclusions regarding non-oil GDP	54

3 The manufacturing sector in Iran — 57

Industrialisation in Iran, 1962–72	58
The post-1973 boom and manufacturing	62

4 Agricultural production in Iran — 73

Land reform and farm structure	74
Prices, costs and production after 1973	87
Conclusions	98

PART III   THE CASE OF NIGERIA

Introduction — 105

5 Macroeconomic adjustment — 107

Oil revenues and changes in aggregate expenditure	107
The pattern of government expenditure	111
The Gross Domestic Product and its components	116

6 Nigerian agriculture in the 1970s

The land tenure system	124
Farming systems and constraints	127
Impact of agricultural policies	135
Prices, production, and incomes	142
Summary	152

7 Conclusions — 155

Statistical Appendix A: Iran	**161**
Statistical Appendix B: Nigeria	**215**
Bibliography	**248**
Index	**255**

Acknowledgements

The author is grateful to several people for their kind and constructive criticisms of earlier drafts of this book. In particular, Michael Lipton, Charles Harvey and David Evans provided me with both criticisms and encouragement. Michael Lipton's detailed commentary on the first draft helped me to synthesise the main message of the book and strengthen the weak points in both the theoretical and the empirical sections, while the institutional emphasis of the book owes very much to Charles Harvey. David Evans provided a critical examination of the Theoretical Part which helped me to better consolidate the empirical findings. Robert Eastwood, who also read the first draft, provided a number of valuable insights and comments which gave me further encouragement and guidance. Others who also read the manuscript and provided comments are Peter Ayer, Masoud Karshenas and V.S. Krishna Moorthy, and the author is grateful to all of them. Thanks also go to Negin Ganjavi for helping to draw the diagrams for the book. The book has its origins in a Ph.D dissertation submitted to the Institute of Development Studies of the University of Sussex. The author is grateful to many of IDS students and staff.

Introduction

This book is an examination of the domestic economic response of two middle-income, oil-exporting countries, Iran and Nigeria, to increases in their foreign exchange earnings during the 1970s and early 1980s. It also evaluates the consequences of the type of adjustment path followed in these two countries. In this sense this is both a positive as well as a normative study. The main argument of the book is that there is no unique 'automatic adjustment mechanism' which necessarily results in an oil economy.[1] The accrual of the oil earnings of these two countries to their central governments brings to the forefront fundamental issues of public policy, and the policies adopted are by no means predetermined. Indeed if this book can contribute to illustrating some of the key issues which face policy-makers in such an economic environment the author will have achieved his purpose.

It should be added that this book is not intended as a rigorous proof of a particular economic model, but rather as an exercise in using economic analysis to guide the investigation in both scope and emphasis, as well as providing a framework for policy analysis. Moreover, the rapidly changing production structure of the economies in question makes modelling on the basis of fixed input–output coefficients a rather dubious exercise, apart from the fact that the input–output model of Iran is only from 1967 and none exists for Nigeria.[2]

The theoretical framework of this book, as presented in Chapter 1, involves three stages of analysis. In stage one it is assumed that there is no government intervention and the macroeconomics of the export boom in terms of changes in absorption and changes in relative output prices is spelled out. This is the so-called 'Dutch-Disease' model which has been used by many economists for analysing the economies of oil-exporting countries. In stage two of the analysis, however, it is argued that the Dutch-Disease model can be usefully complemented by introducing capital inputs (namely, intermediate capital goods and land) into the model for enhancing its explanatory power.

In stage three, government intervention is introduced into the analysis and it is argued that government actions can alter the supply response of the economy, especially at the sectoral level. These three stages of analysis will provide the foundations for a normative discussion of policy choices in an oil-exporting developing country.

In stage one the argument is based upon the recent macroeconomic literature as applied to a small open economy with two sectors. These two sectors consist of a traded goods sector which is exposed to international trade and a non-traded goods or home goods sector which is sheltered from international trade due to transport costs; non-traded goods (or services) are neither exported from nor imported to the country so as to equalise their domestic and their border price. At this stage only transport costs are the cause of a product being non-traded since the assumption is that no government intervention is present. The prices of traded goods are given externally through international trade while the prices of non-traded goods are determined internally according to the supply and demand situation in the country. The classic examples of non-traded goods (or services) are haircuts and houses.

With the export boom (due to the rise in the price of oil), assuming the economy is previously in equilibrium, the Dutch-Disease model predicts a shift in relative prices in favour of non-traded goods. This is because, with the export boom, the supply of foreign currency rises relative to the supply of domestic currency, pushing down the price of the former in terms of the latter, so that foreign exchange becomes cheaper and there is an increase in demand for imports. The greater demand for imports results in the decline of the accumulated foreign exchange reserves which originally resulted from the export boom. Cheaper imports imply a fall in the relative price of traded goods in the country. The rise in the relative price of non-traded goods would be expected to result in higher production of non-traded goods and lower production of domestic traded goods which would be substituted by imports. This is because the higher price of non-traded goods would mean a higher real wage in terms of traded goods and thereby a fall in output and employment in that sector and a lower real wage in terms of the non-traded goods and a rise in employment and output there. In other words, there would be a shift in labour away from traded and towards the non-traded sector resulting, given full employment, in fall in the output of the non-oil traded sector.

The export boom also has substantial income effects resulting in higher levels of expenditure on both traded and non-traded goods. Greater expenditures on traded goods results in more imports and a decline in the accumulated foreign reserves. The combination of higher levels of expenditure and changes in relative prices brings about 'adjustment' or both external and internal balance: external balance through a greater level of imports and internal balance through higher prices and thereby equilibrium in the market for non-traded goods. The analysis is carried out in terms of the current account and then extended in order also to cover some of the monetary aspects of the adjustment process.

The lower production of traded goods at home arises from the lower profitability of producing such goods given the relative fall in their output price and the consequent rise in the real wage in terms of traded goods. Profitability, however, is affected not only by output price relative to wage but also by capital input prices. Distinguishing between traded and non-traded capital inputs, the composition of such inputs in terms of being traded and non-traded would also have a direct impact on profitability in each sector during the adjustment process. The higher the share of non-traded capital inputs in the cost of production of a sector, the higher would be the adverse impact of real appreciation on value added in that sector. The explicit introduction of the role of capital inputs and their prices would then constitute the second stage of the analysis. Within the industrial sector, for example, with the rise in the relative price of non-traded goods, *ceteris paribus*, those industries whose inputs are largely based on domestic resources will be more pressurised than those with a larger component of imported inputs. This is because domestic resources are likely to contain a greater element of non-traded goods which would now be more costly than imported intermediate inputs.

In stage three of the analysis the impact of government intervention on the output of each sector is explored. A number of hypotheses concerning how the government is likely to behave in boom conditions are proposed. Such hypotheses concerning government behaviour should, however, not be viewed as cast in stone since deviation from expected behaviour is not only possible but in many cases desirable. Indeed the analysis seeks to show what government could avoid when managing large external transfers of resources from abroad. These hypotheses will be mentioned shortly.

In discussing government intervention and its impact on the economy two pairs of distinctions suggest themselves, and they have been used in the following chapters. The main distinctions are between the 'planning mode' and the 'crisis mode' of government intervention and between the 'form' and the 'content' of government intervention.

Interventions within the 'planning mode' normally reflect a 'development strategy' and are usually long-term in nature. Interventions in the 'crisis mode', on the other hand, are usually *ad hoc* in nature and stirred by earlier developments. Such crisis-mode interventions are usually short-term in nature and in many instances can contradict the objectives and methods of the interventions in the planning mode. It is argued in this book that the mode of government intervention (planning or crisis) affects the output of the economy in both the traded and non-traded goods sectors and should therefore be considered as a factor when analysing output response to external shocks.

Apart from the above modes of government intervention, one can distinguish between the 'form' and the 'content' of government intervention. The 'form' here refers to the procedure that the government uses to arrive at its policy decisions, for example the spending appraisal method adopted (or not) for deciding among alternative spending proposals. The 'content' of intervention here refers to the type of policy adopted. In this regard the principal areas explored in this thesis are: fiscal and monetary policy; sectoral allocation of investment policy; land-rights policy; credit policy; pricing policy for food crops; tariff policy; and consumer-subsidy policy.

At the macro level, the pace of increase in domestic expenditure is set by the growth in money, assuming velocity does not change in the opposite direction by the same amount. The growth in money is partly a function of the government budgetary expenditures in oil-exporting countries. Changes in money can be regarded as indicating changes in aggregate expenditure. The composition of aggregate expenditure would depend upon the changes in relative prices of traded and non-traded goods. Therefore, from changes in government expenditure and its monetary consequences, one can derive the expected changes in relative prices of traded and non-traded goods.

At the sectoral level, the land-rights policy affects especially the agricultural output by affecting both the incentives to invest on the land and access to credit by small farmers; credit policy affects availability, price, and access to both technology and working capital, hence affecting the level of output; tariff policy affects incentives in the domestic traded goods industry and thereby their level of output; pricing policy for food (also export) crops affects the respective supplies of these sectors and in the oil-economy case the government has additional financial resources to make these policies especially effective. The sectoral allocation of public investment and the types of project undertaken, especially in such sectors as infrastructure, would affect the cost of production, and therefore output, in various sectors of the economy. Finally, consumer subsidies can prevent price rises at the sectoral level and thereby, depending upon the importance of the subsidised sub-sector in the consumer expenditures, subsidies could affect the aggregate level of prices and thereby affect the pace of the real appreciation of the exchange rate.

The direct accrual of oil revenues to the government is likely to increase the role of the government in the economy resulting in an increase in government's direct expenditure, including capital expenditure. Government's capital expenditure is mostly on physical and human infrastructure and on services not usually provided by the private sector. Government's services are non-traded in the sense that they are neither imported to nor exported from the country. Moreover, government's spending on infrastructure only generates non-traded goods in the first round in the above sense, for example construction of schools, dams, or roads. Such infrastructure could, of course, indirectly prove beneficial to the sectors with a traded output but infrastructure as such is non-traded. In other words, the increase in government's revenue and expenditure in an oil-exporting country is likely to result in an increase in investment in what government's capital expenditure usually consists of – non-traded goods. Moreover, with the foreign exchange constraint released after the oil boom, it is likely to observe a shift of public investment towards non-traded goods since they become a key constraint for the realisation of planned targets.

Concerning the sectoral allocation of public investment, this book discusses both the allocation between traded and non-traded goods and allocation within the traded goods sector. The allocation of public investment towards the non-traded goods sector is especially important since it would reinforce the shift of private investments towards the non-traded sector during the boom and this would constitute a double drain of investment resources away from traded goods with implications for output in that sector.

In analysing the allocation of public investment within the traded goods sector, the main question which has been posed in this book is whether the government follows a unimodal or a bimodal strategy of development. A unimodal strategy aims at the progressive modernisation of the entire sector, whereas a bimodal strategy is a crash modernisation programme which concentrates resources in the highly commercialised and usually capital intensive sub-sectors resulting in a skewed pattern of public investment. It should be added that the distinction between the bimodal and unimodal strategy also applies to the non-traded sector, for example rural feeder roads versus modern motorways; but in this book this distinction has been largely limited to the agricultural sector since the purpose is to be suggestive rather than comprehensive. Within the agricultural sector bimodalism could have an adverse impact on output and therefore a high opportunity cost to the economy. This is because such a strategy is likely to marginalise the bulk of the agricultural producers who are small-scale and who would not benefit from a strategy which aims at the large-scale commercialised and capital-intensive sub-sectors in agriculture.

Having briefly mentioned what could happen with government intervention, let us propose a number of hypotheses concerning actual government behaviour which are likely to be ascertained in practice:

1. An increase in 'government-type' public capital expenditure which means largely investments towards the non-traded sectors especially concentrated in physical infrastructure. This increase comes about especially because, with the foreign exchange constraint released, the non-traded goods sector, and especially infrastructure, becomes a key constraint to the rise in investment.
2. A bimodal or skewed pattern of development in the traded sector especially reflected in a shift of public sector credit towards large-scale mechanised sub-sectors in both agriculture and industry; imported machinery substitutes domestic labour and large-scale mechanised operations are viewed as

a faster way to increased production and also a way of bypassing the institutional constraints on developing small-scale production. This rather skewed pattern of investment in directly productive sectors is also likely to involve direct government intervention in production, for example state farms.

3 A general loosening of tariff barriers largely due to the high demand for imports and the ability to pay for them.

4 The maintenance of cheap-food and -energy policies, especially in the urban areas. It should be added that this is not unique to an oil-exporting developing country but it certainly becomes more effective in the boom period of these economies because of the large financial resources available to the government.

5 Increased resort to crisis-mode type of intervention by the government due to the emergence of severe bottlenecks in the economy and the accelerated rate of economic activity.

These hypotheses concerning government behaviour conclude the third stage of the analysis contained in this book. The above three stages of the analysis constitute the 'positive' part of the book. There is also a normative component in this exercise. The economic dynamics of the oil boom plus the likely government behaviour in this situation will lead to a position of disadvantage facing the traded goods sector. This could lead to a loss of diversification in the export base and a decline in import-substituting industry and agriculture. The real appreciation of the exchange-rate – the rising relative price of non-traded goods – occurs within a comparatively short period of time. It is highly unlikely that these countries can transform their traded sectors quickly enough so that they can cope with the period of rising costs due to higher prices of non-traded goods. The economy would then increase its dependence on a single traded resource, namely oil. Oil is both a depletable resource and has a rather erratic world price. With limited oil reserves, dependence on oil creates uncertainty and possible hardship in both the short and long run. The depletion of oil in the long run can also generate the need for radical and swift economic restructuring with high potential costs.

This discussion of how to avoid the possible consequences or unfortunate side-effects of the oil boom can run on two separate grounds. First, one can argue that the best strategy in the long run is to minimise government intervention. In this approach the market forces themselves would signal the correct set of relative prices which would lead to diversification in the traded goods sector in time before the oil runs out. The capital markets would provide funds for the restructuring of the traded goods sector against their future prospects after depletion.

Second, one can argue that the non-intervention policy is neither possible nor desirable. What is required is a set of policies which allow the economy a breathing space for developing a diversified resource base in order to minimise the impact of depletion of oil in the long run and of oil price fluctuations in the short run.

Commenting on the two approaches, it can first be said that the non-intervention approach assumes that the private investment will readjust from oil and non-traded goods to non-oil traded goods in good time by correctly predicting depletion, and invest accordingly through funds from the capital markets. The capital markets in most oil-exporting developing countries, however, are poorly developed. The government usually needs to guarantee various stocks before they are purchased by the public and this would defeat the very purpose of the non-intervention approach. The imperfection of the capital markets implies that investment is likely to shift in large part towards the non-traded sector, reflecting private profitability under oil-boom conditions. If the economy is not prepared to deal with the slowdown and end of oil revenues, at the end of the boom the country would be faced with a massive non-traded sector which is not functional due to the shortage of foreign exchange. The irreversibility of investment in the non-traded sector as, for example, in the construction of motorways or high-rise buildings means that it is not very meaningful to argue that by relying on the market forces, while the oil revenues near exhaustion, the economy can shift resources back into the traded sector. This is because by that time the resources are already sunk and such investment has little salvage value.

Another point which can also be made against the reliance-on-market-forces approach is that development of the traded sector is not something which can be done instantaneously and requires decades of investment, and especially learning-by-doing. This means that without policies to encourage the development of the traded sector, the sector would suffer from a considerable knowledge gap at a time when it becomes competitive again after oil revenues decline and the price of non-traded goods falls. In other words, there are fundamental externalities involved in the pattern of expenditure adopted during the oil boom and reliance on the private sector with short-term profitability goals is unlikely to meet the long-term human and physical resource-development requirements of the country.

The normative argument adopted in this book therefore, runs along the second approach concerning resource allocation in the oil economy, the argument being that the cost of adjustment after depletion in the future as well as the impact of short-term fluctuations on the economy can be considerably reduced with a correct set of policies which by no means exclude a major role for both the market and the private sector. What is required is a spending programme which, first, in the aggregate does not result in a sudden rise in the relative price of non-traded goods and the real appreciation of the exchange rate takes place at such a pace as to allow for the domestic-resource-based and efficient industries to survive; second, does not result in a sharp increase in public investments in the non-traded sector so that the traded sector is not crowded out in terms of public investment; and third, includes specific productivity-enhancing and market-development policies for the traded sector in in general and small-holder agriculture in particular; and fourth, relects a unimodal pattern of development in line with the evolution of existing institutions and technologies and does not require their sudden transformation. In sum, the main problem facing an oil economy is how to raise total factor productivity so that the economy can diversify away from dependence on oil.

This was an outline of the main arguments (both positive and normative) of the book. These issues are explored both theoretically and empirically in the following chapters. Chapter 1 is the theoretical presentation of the general framework which motivates the book in scope and emphasis. Chapters 2, 3 and 4 are an empirical investigation of the adjustment policies and consequences in Iran, with particular emphasis on the implications of the adjustment policies on the industrial and the agricultural sectors of the economy. Chapters 5 and 6 are a similar exercise for Nigeria. Chapter 7 concludes the book and highlights policy issues arising from the analysis.

Notes

1. In the literature concerning adjustment in oil-exporting countries it is not uncommon to find that the discussion is conducted in terms of an inevitable or unique character of the type of adjustment process followed in such economies. Phrases such as 'automatic adjustment mechanism' or the 'Dutch-Disease Syndrome' are often used, implying the lack of policy choices and the uniqueness of the process of adjustment.

2. For an attempt to put the Nigerian economy into a general equilibrium model see Taylor.

PART I
THEORY

1 The theoretical framework

The theoretical framework of this book draws largely from three main sources. The first is recent developments in the theory of international trade concerned with the general equilibrium effects of a boom in a sector of the economy producing traded goods. This provides the underlying economic model for understanding the relationship between the oil-financed rise in domestic expenditure, the appreciation of the real exchange-rate, that is, the rise in the relative price of non-traded goods, shifts in the structure of production and consumption, and the re-establishment of internal–external balance. The problem has become known as booming-sector or 'Dutch-Disease' economics.[1]

The second source is the literature on the impact of oil revenues on government policies, programmes and projects and how the accrual of oil income to the government results in a particular pattern of expenditure (see Lewis 1982, Essang 1977).

Finally, Johnston and Kilby's (1975) discussion of 'unimodal' and 'bimodal' patterns of agrarian change provides the third source of the underlying framework for this book.

This chapter begins by introducing non-traded goods and then proceeds by an analysis of internal–external balance and how the oil shock initiates changes in absorption, relative prices, and output. For simplicity, the economy is divided into a traded and a non-traded sector, and the main question is how higher levels of expenditure and changes in the price of non-traded goods bring about internal–external balance. The model is one for a small country with given terms of trade and domestic full employment. The change in the relative price of traded to non-traded goods brings about change in the sectoral real product wage which in turn causes labour reallocation across the sectors, thereby affecting the level of output in both the traded and non-traded sectors.

The model is then extended to include money as a 'determinant', or rather an indicator, of aggregate expenditure in the sense that excess money balances are spent. Through its influence on aggregate spending, money affects the level of prices, relative prices, and the balance of payments. The monetary model adopted assumes no sterilisation on the part of the monetary authorities and a fixed exchange rate. In fact, at this stage, earlier referred to as stage one of the analysis, the model is one of an automatic adjustment mechanism otherwise known as a price species–flow mechanism.

The chapter then goes on to consider how the above analysis can be complemented through the introduction of the role of traded and non-traded capital inputs. For example, the real appreciation of the exchange rate has a differential impact on the various sub-sectors of the traded goods sector largely depending upon the composition of inputs in these sub-sectors in terms of being traded or non-traded and, more fundamentally, depending upon the ratio of non-traded capital inputs to the value added in each industry.

There then follows an investigation in general terms of some of the public policy aspects of the adjustment process. Here, attention is drawn to the role of government policies as they affect the macroeconomic as well as sectoral adjustment in an oil-exporting country. Here a key point of the book is once again emphasised, namely that the Dutch-Disease model by itself is insufficient to explain supply response, and the role of government policy must be explicitly considered if one is adequately to explain sectorial output changes during the oil boom. In this sense the book can be regarded as a critique of the concept of an 'oil economy' which misleadingly implies determinate results and inevitable outcomes consequent to an oil boom. Moreover, the impact of large revenues in a short period of time on the scope and character of government intervention in the economy, especially in programmes and projects in the agricultural sector, is examined.

Non-traded goods

Consider an economy in which total production and total expenditure can be divided into two categories: traded goods and non-traded goods. The distinction broadly corresponds to sheltered and unsheltered industries. Traded goods are those with prices determined in world markets. They consist of exported and imported goods. Non-traded goods are those which do not enter world trade either due to high bulk and low value, hence not being profitable to trade, or due to commercial controls by the government which would prohibit or limit trade in that commodity. In other words, at a given exchange rate the prices of non-traded goods would differ substantially across countries.

Transport costs provide a convenient explanation for a commodity being internationally non-traded and they are used here for discussing non-traded goods in theoretical terms.

Assume a commodity has a given world P_f and assume the commodity is consumed and produced at home. The question is when this commodity will be imported, exported, or non-traded.

In Figure 1.1 let Q represent *ad valorem* transport costs, N a fixed percentage charge per unit value. The domestic and the foreign price are shown on the horizontal and vertical axes as well as a cone that defines the region of non-traded goods which will now be derived.

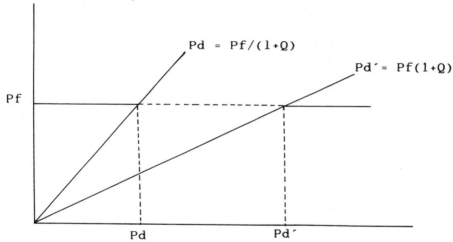

Figure 1.1

For the commodity under discussion to be exported, domestic price inclusive of transport costs $P_d(1 + Q)$ would have to be smaller than or equal to foreign price. At $P_d = P_f/(1 + Q)$ the commodity is at the margin of being non-traded and exported. For the commodity to be imported, the foreign price inclusive of transport charges $P_f(1 + Q)$ would have to be smaller than or equal to the domestic price. This defines the maximum price for which the good is non-traded before it becomes imported: $P_{d'} = P_f(1 + Q)$.

Figure 1.1 shows the world price P_f and the corresponding maximum and minimum prices that can prevail in the home country. The horizontal distance between the two prices P_d and Pd' defines the range of prices for which the commodity is non-traded. It is sheltered from the world market competition by transport costs. If the domestic price becomes borderline, the commodity becomes traded at the transport-cost-adjusted world price. In other words, a non-traded good is here defined as any good which is too expensive to import and not cheap enough to export. The analysis also suggests that an expansion in domestic demand for the commodity which raises domestic price will push some goods to the margin of becoming imported. Conversely a fall in demand and prices will make some non-traded goods exported. Moreover, an import tariff in the home country at rate t would increase the range of prices at which the commodity is non-traded by $P_{d'} = P_f(1 + Q)(t + 1)$ (more on the tariff issue when government is introduced into the analysis below).

The core adjustment model

The model discussed here has its origins in the writings of Wilson (1931), Swan (1960), and Salter (1959) in the Australian trade literature. The summary below is based on Dornbusch (1980).

The country is assumed to be a price taker in the world market for imported and exported goods alike. The terms of trade are exogenously given, and a distinction between importables and exportables becomes unnecessary for the main process to be discussed. Importables and exportables are thus aggregated into a composite commodity called 'traded goods', as distinguished from 'non-traded goods'. The oil economy would thus be treated as any small open economy. The oil revenues are incorporated into the model as net transfers from abroad and the oil sector itself, being enclave in nature, does not feature in the model. The other assumption of the model is domestic wage and price flexibility. This would ensure full employment. The analysis proceeds by developing first, the supply side of the model, second the demand conditions, and third internal–external balance and the analysis of disequilibrium.

Factor markets and output supply

The two sectors, traded and non-traded goods, are assumed to have specific capital in a fixed amount that is immobile between sectors in the short run.[2]

The labour force is fixed in the aggregate but the allocation between sectors is endogenous. Given the fixed capital stocks in each industry and perfectly competitive markets, the demand for labour in each sector can be derived as a function of technology, the capital stock, and the product wage (product wage is the wage measured in terms of the output of the respective sectors). Equilibrium in the labour market then requires:

$$L_t(W/P_t, K_t) + L_n(W/P_n, K_n) = L \tag{1}$$

where W is the nominal wage, P_t and P_n are traded and non-traded goods prices, and K_i is the capital stock in the ith sector ($i = n, t$). The aggregate labour force is equal to $L = L_t + L_n$. With diminishing returns, the demand for labour will be a function in each sector of the product wage, or wage defined in terms of the output of each sector.

The labour-market equilibrium condition for the equilibrium wage rate can be solved by putting the wage rate as a function of output prices and capital endowments:

$$W = W(P_n, P_t; K_i) \tag{2}$$

The equilibrium wage is a linear homogeneous function of output prices. Specifically, for given capital stocks and productivity we have:

$$W\sim = ZP_n\sim + (1-Z)P_t\sim - Z(P_t\sim - P_n\sim) \tag{3}$$

where ~ denotes a percentage change and Z, which can take any value between 0 and 1, denotes the share of expenditure on traded goods. Equation (3) is derived from Equation (1) by differentiation.[3]

Equation (3) is a key relation. It states that the change in the equilibrium wage rate is equal to the change in the price of traded goods adjusted for changes in the relative price of traded goods in terms of non-traded goods.

Alternatively, taking $P_t\sim$ or $P_n\sim$ to the left-hand side, we derive a relation between product wages and relative prices:

$$W\sim - P_t\sim = -Z(P_t\sim - P_n\sim)$$

$$W\sim - P_n\sim = (1-Z)(P_t\sim - P_n\sim) \tag{4}$$

A fall in the relative price of traded goods raises the equilibrium relative wage in terms of traded goods and lowers the relative wage in terms of non-traded goods. In other words, the impact of higher relative price of non-traded goods on the real wage depends upon the share of non-traded goods in the total expenditure of the workers. Given the labour force, a lower real wage in terms of non-traded goods implies a rise in employment there, while the higher real wage in the traded goods sector discourages employment there. Therefore, there is a shift of labour away from traded to non-traded sector and a fall in output of the non-oil traded sector under the pressure of increased labour costs.[4]

Now consider the equilibrium conditions for the supply side. Denote the relative price of traded goods in terms of non-traded goods as v, so $v = P_t/P_n$. From Equation (4) and the demand functions for labour, it is evident that employment and output in each sector is a function of relative prices:

$$Y_t = Y_t(v)$$

$$Y_n = Y_n(v) \tag{5}$$

In Equation (5) output of non-traded goods is an increasing function of the relative price of non-traded goods, and vice versa for traded goods. These are the full-employment supply schedules along the transformation curves in Figure 1.2. The Figure illustrates the transformation curve *CC* between traded and non-traded goods and a relative price ratio with the associated levels of equilibrium output in each industry:

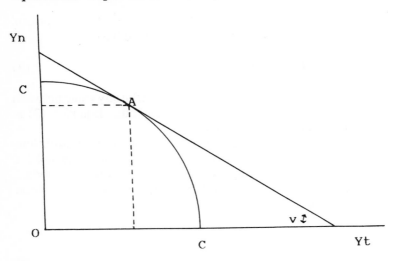

Figure 1.2

Demand conditions

Demand functions for non-traded goods can be written as a function of relative prices v and real expenditure measured in terms of non-traded goods (or any other numeraire):

$$D_n = D_n(v, E)$$

$$D_t = D_t(v, E) \tag{6}$$

where E measures real spending in terms of non-traded goods, or $E = D_n + vD_t$, +b and $D_n(v, E)$ and $D_t(v, E)$ are the demand functions. A rise in the relative price of non-traded goods, given E, reduces the demand for non-traded goods towards traded goods. There is substitution away from non-traded goods towards traded goods. The rise in the relative price of non-traded goods would also have income effects which would reduce demand for non-traded goods. The rise in the relative price of non-traded goods would have an ambiguous effect on the demand for traded goods since the substitution effect raises demand for traded goods, whereas the income effect would reduce that demand. A rise in real spending raises both demands (assuming both traded and non-traded goods are not inferior goods).

Internal and external balance

Combining the supply side and the demand side of the economy, we can show the combinations of expenditure levels and relative prices that will yield internal and external balance. Figure 1.3 shows three schedules.

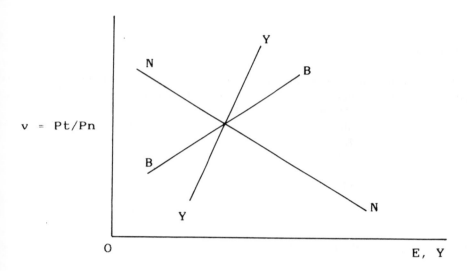

Figure 1.3

The *YY* schedule shows the value of total output in terms of non-traded goods for each relative price, corresponding to the vertical intercept of the price line in Figure 1.2. The schedule is formally defined thus:

$$Y = Y_n + vY_t \tag{7}$$

Along the *BB* schedule we have external balance, that is, $Y_t - D_t = 0$. A fall in the relative price of traded goods increases demand but lowers supply of traded goods and accordingly there is a trade deficit. To eliminate the trade deficit, real spending and therefore demand for traded goods must fall. Thus the *BB* schedule is positively sloped. As P_t/P_n falls, we move from AB to BA and hence from trade surplus to trade deficit and accordingly real spending must fall (keeping the spending on non-traded goods constant by definition since D_n is being used as the numeraire).

Along the *NN* schedule we have the non-traded goods market equilibrium. Assuming that the substitution effects dominate the income effects of a relative price change, the schedule is drawn with a negative slope. A fall in the relative price of traded goods reduces demand and creates an excess supply of non-traded goods that has to be offset by an increased level of spending to maintain equilibrium in that market.

The equations of the *BB* and the *NN* schedules are, respectively:

$$Y_t(v) = D_t(v, E) \tag{8}$$

$$Y_n(v) = D_n(v, E) \tag{9}$$

Disequilibrium

Disequilibrium can now be characterised as overspending or underspending and in terms of overvalued or undervalued real exchange rate, the real exchange rate being defined here as the relative price of traded goods *vis-à-vis* non-traded goods – the justification of defining the real exchange-rate in this way will be given shortly.

The excess of income over spending is equal to:

$$Y - E = v(Y_t - D_t) + (Y_n - D_n) \tag{10}$$

Or, writing the expression in terms of the trade balance $Y_t - D_t$:

$$v(Y_t - D_t) = (D_n - Y_n) + (Y - E) \tag{11}$$

Equation (11) states that the trade-balance problems are either a result of imbalance between spending and income or of a disequilibrium in the non-traded goods market. A trade surplus reflects either an excess of income over spending or an excess of demand for non-traded goods. When the non-traded goods market clears, the trade surplus is equal to the excess of income over expenditure.

The meaning of the real exchange rate should now become apparent. This is a definition of the exchange rate in terms of the relative price of two composite commodities, namely traded and non-traded goods, and not of two different national moneys. The advantage of defining this relative price as the real exchange rate is that it allows one to grasp the relative price adjustment required for achieving internal–external balance. As an example, suppose that the nominal price of non-traded goods were given. Then adjustments in the equilibrium relative price would have to come from changes in the nominal price of traded goods. Given world prices, this means that the nominal exchange rate would have to adjust. In other words, changes in the equilibrium relative price of non-traded goods would have to come only through changes in the nominal exchange rate while the nominal price of non-traded goods remains the same. If changes in the nominal exchange rate are also accompanied by changes in the relative price of non-traded goods in the same direction, however, the net effect on the trade balance may be nil. In other words, there has been no change in the relative price structure between traded and non-traded goods so as to bring about trade balance, that is, the real exchange rate has remained the same.

Adjustment for internal and external balance

Having discussed the structure of the model, the role of income and price effects, and the meaning of a number of key terms, such as non-traded goods and the real exchange rate, we are now equipped to discuss the case of a disturbance in the external balance due to a large transfer of resources from abroad – the rise in oil revenues.

Defining external balance as current account equilibrium, we can note in Figure 1.4 a shift to the right of the YY schedule by the transfer and of the BB schedule by $1/m$ times the transfer, where m is the propensity to spend on traded goods.

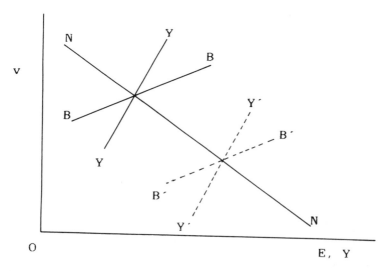

Figure 1.4

The diagram illustrates that the equilibrium relative price of traded goods falls, and income and spending will rise in order to maintain internal and external balance. In other words, the equilibrium price of non-traded goods rises, that is, there is an appreciation of the real exchange rate. The rise in the price of non-traded goods induces a substitution towards traded goods until the home goods market clears, or until the trade deficit equals the transfer.

The analysis of the transfer can also be conducted in terms of demand and supply schedules in the non-traded goods market. It was noted earlier that the supply of non-traded goods, given labour-market equilibrium is an increasing function of the relative price of non-traded goods in terms of traded goods. Figure 1.5 illustrates the situation.

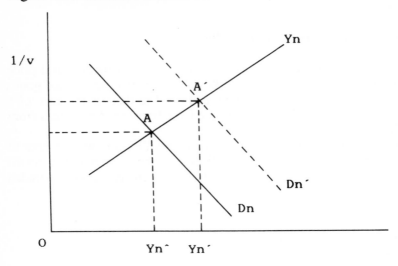

Figure 1.5

Consider the supply schedules Y_n of non-traded goods. The increasing marginal costs of non-traded goods production are a reflection of diminishing returns in the two sectors. With a transfer, disposable income rises, and so does expenditure. Demand for non-traded goods rises at each relative price, thereby shifting out the demand schedule by the marginal propensity to spend on non-traded goods times the transfer. The equilibrium relative price of non-traded goods rises, and so does the level of non-traded goods production.

What factors are responsible for the correspondence between worsening trade balance and the transfer? In other words, what exactly brings about 'adjustment'? In part, the trade balance deteriorates because of the increase in demand for traded goods which increases as a direct function of disposable income. For the rest, the trade balance deteriorates because, on the supply side, resources are withdrawn from traded goods production in order to supply increased quantities of non-traded goods. On the demand site, the rise in the relative price of non-traded goods leads to substitution towards traded goods and a consequent increase in the trade deficit. In other words, there are income and substitution effects on the demand side and substitution effects on the supply side which all contribute to bringing about the fall in external reserves and a worsening of the trade balance.

Factor incomes adjustment

Given the fixed capital stock and perfectly competitive markets, the demand for labour in each sector depends on the wage – price relationship in that sector (Equation 1). In other words, the demand for labour depends negatively on the sectoral real product wage or W/P_i where W is the wage and P_i the relative price of sector i. This means that the rise in the relative price of one sector by more than the wage would increase employment and thereby output in that sector. At full employment, increased output of sector i would involve higher cost per unit since it would come about through an increase in that sector's labour/capital ratio which means falling sectoral marginal physical product of labour. The marginal physical product of labour in sector i (MPP_i) falls because more labour (reallocated from the other sector) is being applied to a given amount of capital. The mobility of labour and the immobility of capital (due to being sector-specific) means that a change in the relative price of traded to non-traded goods would have an uneven impact on factor incomes. Labour mobility allows labour to shift out of the traded sector and thereby maintain or raise its real income, while capital immobility has the opposite consequence. Immobility means that returns to capital fall sharply in the sector with the lower relative price while returns to capital in the sector with a higher relative price rise sharply.

To see this argument more clearly, let R_n denote the rental on capital in the non-traded sector, P_n the relative price of non-traded goods, and MPP_n^k the marginal physical product of capital in the non-traded sector. Thus we have:

$$R_n = P_n \cdot MPP_{kn} \qquad (12a)$$

since rental on capital must match the value of capital's marginal product. With a rise in P_n, MPP_{kn} also goes up since more labourers are added to a given number of machines. It follows that if P_n goes up, then R_n must go up by more than P_n, that is $R_n \sim > P_n \sim$. The opposite holds for the traded sector. Let R_t denote rental on capital in the traded sector, P_t the price of traded goods, and MPP_t^k the marginal physical product of capital in the traded goods sector. Then:

$$R_t = P_t \cdot MPP_{kt} \qquad (12b)$$

With a fall in P_t, labour moves out of the traded sector and thus fewer workers are working with a given number of machines. Therefore, MPP_t^k must fall. It follows that with a fall in P_t, R_t must fall by a greater amount, that is, $-R_t \sim > -P_t \sim$ or $P_t \sim > R_t \sim$.

Now consider what happens to the wage. Let W denote the wage rate, P_n the relative price of non-traded goods, and MMP_{ln} the marginal physical product of labour in the non-traded goods sector. Then:

$$W = P_n \cdot MPP_{ln} \tag{13}$$

Equation (13) states that the wage rate must equal the value of labour's marginal physical product. With the rise in the relative price of non-traded goods more workers are hired in that sector since W/P_n falls, which also means that MPP_{ln} has to go down as the result of more workers being applied to the same amount of capital. The opposite takes place in the traded goods sector. The movement of labour from the traded to the non-traded sector, the consequent drop in MPP_{ln} and the rise in MPP_{lk} means that the nominal wage rate is equalised for both sectors. Moreover, the real wage falls in terms of non-traded goods and rises in terms of traded goods, that is, $P_n \sim> W \sim> P_t \sim$. Having established earlier that $R_n \sim > P_n \sim$ and $P_t \sim > R_t \sim$, it follows that with a rise in the relative price of non-traded goods, that is, an appreciation of the real exchange rate, we have:

$$R_n \sim> P_n \sim> W \sim> P_t \sim> R_t \sim \tag{14}$$

Thus the model predicts an unambiguous result for the specific factor rents which rise for the non-traded goods sector in terms of both prices and fall for the traded goods sector in terms of both prices. The rise in the return to capital in non-traded goods sector will be proportionately greater than the rise in the price of non-traded goods induced by the real appreciation. In other words, the effect of the product price change is magnified in its incidence on return to capital specific to non-traded sector.[5]

Moreover, the impact of the rise in the relative price of non-traded goods on workers' welfare would depend upon the weight of traded and non-traded goods in their consumption pattern. If traded goods make up the bulk of consumer goods, then the higher relative price of non-traded goods and the lower relative price of traded goods should increase the real wage and, therefore, the welfare of the workers.

Money, relative prices, and payments adjustment

Now let us introduce financial considerations into the model, which has so far been developed purely from the real side. Here we shall introduce money as the only financial asset and assume full employment with domestic price flexibility. Again the automatic payments adjustment mechanism in a monetary fixed exchange rate economy is assumed.

We begin with the equality:

$$E = VH \tag{15}$$

where E is nominal spending, H is nominal money holding and V is expenditure velocity. Further:

$$I = PY \tag{16}$$

where I is nominal income, P is the price level, and Y is the level of output.

For the private sector, the rate of increase of money balances is equal to the excess of income over spending, or:

$$H\sim = I - VH \tag{17}$$

where $\sim$ denotes percentage change and change in money balances is a function of the level of money holdings relative to income.

Having introduced money as the determinant of aggregate expenditure in the sense that expenditure comes about as the result of disposing of excess cash balances, we can now introduce the composition of aggregate expenditure between traded and non-traded goods and see how changes in the money stock affect their relative prices. The supply of traded goods and non-traded goods Y_t and Y_n, respectively depends on relative prices:

$$Y_n = Y_n (P_n/P_t)$$

$$Y_t = Y_t(P_t/P_n) \tag{18}$$

The composition of spending depends on relative prices. Assuming constant expenditure shares Z and $1-Z$ for non-traded goods and traded goods, expenditure on non-traded goods and traded goods respectively can be expressed as:

$$P_n D_n = ZVH$$

$$P_t D_t = (1 - Z)VH \tag{19}$$

Equilibrium in the non-traded goods market requires that demand equals supply:

$$Y_n(P_n/P_t) = D_n = ZVH/P_n \tag{20}$$

The equilibrium price of non-traded goods would therefore be:

$$P_N = P_N(H, P_t) \tag{21}$$

since Z and V are constants.

The equation states that the equilibrium price of non-traded goods is a function of the level of money holdings and the price of traded goods. Figure 1.6 illustrates the relationship.

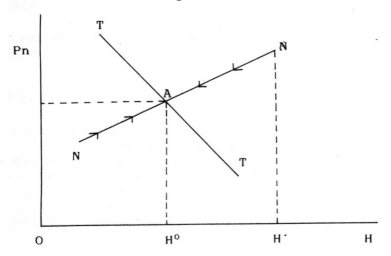

Figure 1.6

NN shows the schedule along which the non-traded goods market clears. The schedule is drawn for a given price of traded goods (or a given exchange rate). An increase in nominal money raises non-traded goods spending and creates an excess demand. To restore equilibrium, the price of non-traded goods must rise. A rise in the price of non-traded goods means a rise in their relative price, given that the price of traded goods is constant. This relative price change would restore equilibrium. The relative price rise leads to reduction in the real value of spending and thus lower demand, and increased supply of non-traded goods.

The gradient of the ray through the origin representing the NN schedule is less the 45° since the price of non-traded goods does not increase by as much as the rise in money. The slope of the NN schedule is determined by the elasticity of the supply of non-traded goods. The higher the elasticity of supply, the flatter the NN schedule.

The trade-balance equilibrium schedule is TT. Along that schedule we have:

$$Y_t(P_t/P_n) = (1 - Z) VHP_t \qquad (22)$$

The schedule is negatively sloped since an increase in the price of non-traded goods would at full employment reduce the supply of traded goods and would require a decline in nominal money and demand to maintain balanced trade. Above the *TT* schedule the trade balance is in deficit, below it it is in surplus.

A monetary disturbance

Suppose now that the money stock rose from H'' to H' (Figure 1.6). At H'' we have equilibrium for both traded and non-traded goods. At H' the non-traded goods market is no longer in equilibrium if the price remains at P_n. The higher level of money and spending implies excess demand for home goods and as a result home goods prices will rise until the point A' is reached at which internal balance is restored. At A', however, there will be a deficit in the trade balance and the money stock would have to fall. There will be a move down to the *NN* schedule until we reach back to the point A.

The case of a transfer (or rise in oil revenues)

The oil price rise and the consequent rise in revenues which is equivalent to a large transfer of resources from abroad can be conveniently treated in this model as a temporary rise in the productivity of the traded sector (Figure 1.7).

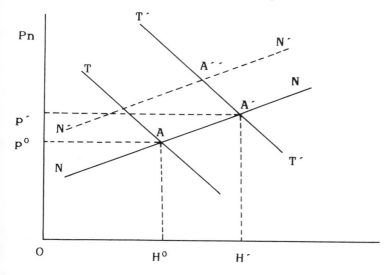

Figure 1.7

TT and *NN* are the schedules along which traded and non-traded goods market clear and A is the point where both traded and non-traded goods markets clear. With a transfer, the *TT* schedule shifts out and to the right, since given relative prices, higher money and spending is required to maintain external balance. Higher levels of money and spending would also require a rise in the equilibrium price of non-traded goods. It should be noted, however, that the difference between the case of a transfer and one of actual rise in the productivity of the traded sector is that in the latter case there is a reduction in the supply of non-traded goods due to competition for labour, in which case the NN schedule would shift up and to the left as shown by $N'N'$ and the new equilibrium would be A''. With a transfer, however, there would only be a shift in the trade-balance equilibrium schedule (equivalent to a rise in productivity of the traded sector per unit of non-traded goods) since the greater supply of traded goods does not come at the expense of non-traded goods. The transfer would result in higher levels of money holdings and higher levels of equilibrium non-traded goods prices without a reduction in their supply.

The new equilibrium is at point A' with a higher level of money stock H' and higher prices of non-traded goods p'. The higher levels of money stock and higher prices of non-traded goods are maintained as long as the shift in TT can be sustained by the higher revenues. This is why earlier higher oil revenues were identified with a *temporary* rise in the productivity of the traded sector. At A' there is also a lower level of domestic production of traded goods due to the higher cost of production resulting from the rise in the prices of non-traded goods. The lower domestic production of traded goods is then substituted through imports which can be paid for by the oil revenues.

This concludes the first stage of the analysis of the impact of an external shock on relative prices and output in the economy. The analysis has so far relied entirely upon the impact of relative *output* prices on resource movement in the economy. Now let us briefly discuss how the introduction of *input* prices would affect or modify the conclusions derived in this first stage.

The role of capital inputs and the cost of production

The discussion so far has been only in terms of output price relative to the price of labour, while the price of capital as such has not entered the explanation of output behaviour. A convenient way of introducing the price of capital into the analysis is to distinguish between traded and non-traded capital inputs and consider how the rise in the price of non-traded goods (and factors) affects the cost of production and value added in each industry. In a typical developing country, traded capital inputs would mainly consist of imported intermediate commodities such as machinery, while non-traded capital inputs would mostly consist of land but also of domestic raw materials. A rise in the relative price of non-traded goods (and factors) would raise the cost of production for an activity which uses land intensively, such as construction. Similarly, a fall in the relative price of traded goods (and factors) would lower the cost of production in the sectors which rely heavily on imported inputs, for example import-substituting manufacture.

Going back to our model with three factors and two commodities and distinguishing between traded and non-traded capital inputs, we arrive at four possibilities.

1. labour combined with imported inputs producing traded goods;
2. labour combined with land producing traded goods;
3. labour combined with imported inputs producing non-traded goods;
4. labour combined with land producing non-traded goods.

Case 1 refers to the domestic manufacturing sector, which is non-protected; case 2 characterises the agricultural sector especially with respect to export crops but also with respect to traded food crops; case 3 refers to the protected manufacturing sector mostly producing import-substitutes; case 4 characterises the construction sector and also non-traded low value or perishable food crops in agriculture. The main point to be emphasised is that, in analysing supply response, traded and non-traded capital inputs should also be considered as they affect value added in each industry. For example, a traded goods sector relying on traded inputs would enjoy some offsetting cost reductions given real appreciation of the exchange rate; its value added would fall by less than if non-traded inputs had a large share in total cost. In the construction sector, for example, although the price of construction and thus profits in construction rise with real appreciation, given the importance of land costs in total costs, higher land prices (since land is non-traded) are likely to lower profits.

To see the impact of change in the price of non-traded goods (after the oil boom) on value added in different activities and how the presence of traded and non-traded capital inputs affect the value added, let P_p denote the pre-boom price of the final traded good. Assume that the oil boom causes a fall in the relative price of traded goods and the domestic price of the traded good becomes $P_p(1 - t)$, with t representing the relative fall in the price of traded goods, that is, the extent of real appreciation of the exchange rate. Suppose the production process uses a traded intermediate good i with the pre-boom price P_i. Let A_i denote the fixed amount of the non-traded good required per unit of final output. Also assume the production process uses a non-traded input n with the pre-boom price P_n and let A_n denote the fixed amount of the non-traded good per unit of final output. At pre-boom prices, value added per unit of final output is:

$$V_p = P_p - (A_i P_i + A_n P_n) \tag{23}$$

and value added at post-boom prices becomes:

$$V_o = P_p(1 - t) - (A_i P_i)(1 - t) - A_n P_n \tag{24}$$

t is the same for both the output and the input since the real appreciation affects all traded goods (and factors) equally in terms of relative price fall. The relative change in value added between the pre-boom and the post-boom period can be written as:

$$VAC = (V_o - V_p)/V_p \tag{25}$$

substituting from Equations (23) and (24) into Equation (25), we have:

$$VAC = -t\,(P_p - A_i P_i)/V_p = -t\,(V_p + A_n P_n)/V_p \tag{26}$$

dividing by V_p and letting $Q_n = A_n P_n/V_p$, that is, Q_n is the ratio of the value of non-traded inputs per unit of final output to value added of the final traded good, we have:

$$VAC = -t(1 + Q_n) \tag{27}$$

Equation (27) establishes an inverse relationship between the share of non-traded inputs in the value of final output and change in value added. Interestingly, Equation (27) also establishes that what is crucial is the share of non-traded inputs in the total cost of production in determining the extent of change in value added after real appreciation of the exchange rate, since traded inputs as such do not affect the value added and they only matter to the extent that they substitute for non-traded inputs. In other words, in an industry with no non-traded inputs, a 10 per cent appreciation of the real exchange rate would lower value added by only 10 per cent, while if the industry also uses non-traded inputs the fall in output price would have a magnified effect on the fall in value added.

For example, assume initially that in two different activities, neither of which use non-traded inputs, price is 200 and cost of traded inputs is 100. Value added would therefore be 100. Assume that we have a 50 per cent appreciation which leads to a fall in output price to 100 and a fall in the cost of traded inputs to 50. Value added would therefore fall to 50 – value added would fall by exactly 50 per cent, which is the same as the fall in output price.

Now assume that, prior to the oil boom, these industries begin to use non-traded capital inputs in addition to traded inputs so that the cost of traded inputs is 60 in the first and 30 in the second, and the cost of non-traded inputs is 40 in the first and 70 in the second:

$$200 = 60 + 40 + 100 \tag{a}$$
$$200 = 30 + 70 + 100 \tag{b}$$

Assume that we have a 50 per cent appreciation which leads to a fall in output price to 100 and a fall in cost of traded inputs to 30 in the first case and 15 in the second case. The new equations are:

$$100 = 30 + 40 + 30 \qquad (a)$$
$$100 = 15 + 70 + 15 \qquad (b)$$

Value added therefore drops by 70 per cent in the first case and by 85 per cent in the second case. Clearly the adverse impact of appreciation on value added is greater for industries with larger share of non-traded inputs in the value of final output.

Two other considerations are relevant before we end the discussion on the cost of production in the traded goods sector and the impact on its output: first, the industrial structure, and, second, the cost of energy.

Concerning industrial structure, it can be said that if these sectors were operating with a high profit margin due to monopoly or protection prior to the relative price changes, then falling relative price would only lower returns, but not necessarily, output, of these industries. Also there is likely to be greater demand for the products of these industries due to the income effects. On the other hand, if prior to the oil boom traded industries were operating at the point of the quality of price and marginal cost, then the simple output as a function of price relationship would hold. What is likely to be observed empirically is the presence of two types of output behaviour regarding price. The competitive smaller industries are likely to behave as though they were operating with price equal to marginal cost, that is, in a perfectly competitive fashion.

The other factor which affects the cost of production is the price of energy. The rise in the price of oil should theoretically also raise the cost of energy in the oil-exporting country itself. This would induce a movement away from industries which are heavy users of energy including oil. This impact, however, will not figure in our analysis simply because, in both countries to be discussed, oil prices at home were heavily subsidised and the oil price rise did not affect the domestic economy via higher oil prices directly.

This brings us to the conclusion of the second stage of the analysis. The general conclusion at this stage is that the introduction of input price changes – traded and non-traded inputs – can significantly affect the actual price and output changes in the traded and non-traded sectors.

Now let us proceed to the third stage of the analysis where the role of government and its impact on the adjustment process is discussed.

The role of government

Since oil revenues accrue directly to the central government in oil-exporting economies or at least the government gets a major share of the revenues through taxation, it becomes critical to look at how government spends this money and at the impact of such expenditure on the domestic economy. In discussing government intervention and its impact on the economy a number of distinctions suggest themselves. These distinctions refer to the *modes* of government intervention and the *types* of government intervention.

Broadly speaking, two modes of government intervention can be distinguished – the 'planning mode' and the 'crisis mode' – although there is overlap between the two. The planning-mode intervention reflects the requirements of the development strategy and economic growth path followed by the government. In the planning mode the state can act as a regulator, spender, lender, and stimulator of private expenditure in different directions it grants to the private sector. Planning-mode interventions are usually long-term in nature.

Crisis-mode interventions are usually *ad hoc* in nature and are stirred up by earlier developments. Such crisis-mode interventions are usually short-term in nature and in many instances contradict the objectives and methods of the planning mode since they could involve sudden reversals of policy. Crisis mode interventions are particularly prevalent in economies undergoing external shocks or economies undergoing an accelerated rate of change, as is the case with oil economies with large injections of resources into the domestic economy in a short period of time.

The importance of the modes of government intervention arises from their impact upon the stability and continuity of policies and hence on output and investment in the economy as a whole. Planning-mode interventions are more likely to have a longer time-horizon and create a more stable policy environment for both the public and private sector. Crisis-mode interventions are likely to lower business confidence, lower the time-horizon of private investment due to greater uncertainty thus generated, and result in poor completion of public projects.

Apart from the mode of government interventions, one can also refer to the type of government intervention which also has significance for the analysis of output supply. The type of government intervention can be analysed both in terms of its form and its content.

The form would refer to how the policy is brought about and implemented. For example, in the case of government expenditure, the form would refer to how the spending is allocated and implemented. Here one needs to consider the investment-appraisal method adopted for deciding among alternative investment proposals. With large financial resources, the existing development plans are likely to be rapidly expanded with little consideration of their feasibility and/or sustainability, especially if skilled personnel for project appraisal are in short supply. With budgetary constraints relaxed, many projects which were not acceptable before become acceptable now. With large budgetary allocations, the bureaucratic imperative of spending the budget results in a hurried approach to planning and implementing public expenditure and an abandonment of sound appraisal methods altogether. National plans are expanded overnight and sectoral growth targets are multiplied arbitrarily. This would also result in poor implementation of many projects, with many being started but few completed and even fewer maintained, especially if the boom is short-lived.

Rent-seeking from public positions becomes the active cause of public sector budgetary allocations. The distribution of public money also becomes a way of strengthening old class alliances and establishing new ones. The oil revenues become a means of consolidating the power base of the state through employment creation, subsidies, and services, especially in the urban areas where political power is usually contested.

A critical implication of the absence of appropriate appraisal of public projects is that not only are inappropriate projects chosen at the sectoral level, but also public investments are likely to rise over and above the limit beyond which extra investment would involve rapidly increasing incremental capital/output ratio. These limits arise from the shortage of goods the supply of which cannot be expanded in the short run. Such goods include managerial and administrative capacity, physical infrastructure, and labour. In other words, there is a limit to the rate of growth of investment set by all domestic factors of production which are non-traded since they face rapidly rising import cost problems. In terms of Figure 1.1, the non-traded input requirements of public investments can only be imported if their price passes the non-traded range and enters the imported range at the right-hand side of the figure and this would involve rapidly rising price. The rapid rise in investments could generate low or even negative rates of return with high opportunity cost to the economy. The form of government intervention as defined above, therefore, can significantly affect both national and sectoral output and income levels in the economy and must be considered as a factor in the adjustment process.

The content of intervention here refers to the kinds of policy adopted. The principal areas of concern in this thesis are: fiscal and monetary policy; sectoral investment policy; land-ownership policy; credit policy; pricing policy for agriculture; tariff and import-control policy; and consumer subsidies. The order of presentation of these policy areas is of no especial significance.

Fiscal and monetary policy

Having described the basic relative price mechanism at work in the real economy resulting from the changes in the level and composition of expenditure, we now need to inquire into the specific monetary and fiscal channels through which government's foreign exchange earnings bring about an increase in domestic expenditure.

With large financial surpluses in the balance of payments, the government may or may not adjust. Adjustment in this sense means allowing the revenues to be absorbed into the domestic economy with the consequence of affecting relative prices and the growth of various sectors. Non-adjustment would involve sterilising the monetary and fiscal impact of financial surpluses through the purchase of foreign assets, that is, investment abroad. Such purchases would show as capital outflows cancelling the receipts in the current account. This is a key point concerning the role of the government since if the private sector received the oil money, there would be no central control over the decision of whether or not to sterilise these funds.

In choosing the option of adjustment, the government has three main potential instruments. First, the government can induce a direct switching of expenditure towards imports by appreciating the value of its own currency relative to those of its trading partners so that traded goods become cheaper. Switching of expenditure towards imports resulting from real appreciation would eliminate the financial surpluses and bring the current account into balance.

The second possibility is to maintain a fixed exchange rate but increase aggregate expenditure through public spending financed not by domestic taxation but by the oil revenues. In this case, domestic inflation would be more rapid than inflation abroad due to the rise in the relative prices of non-traded goods. Greater domestic expenditure would increase imports, partly due to the rise in income and partly due to the switching of expenditure towards cheaper foreign goods. Again this process would eliminate any financial surpluses in the current account, bringing it into balance.

The third instrument of adjustment, although not usually an independent option, is to reduce commercial barriers to trade such as tariffs and import quotas. Lower prices of imported goods due to the elimination of controls should have a similar expenditure-switching effect to real appreciation, especially if barriers to imports are initially high. If such controls are not large enough, however, reducing commercial controls on imports may be insufficient as an independent adjustment option and it would have to be used in concert with the other two options.

Considering the experience of a number of less developed oil-producing countries, the option of the explicit appreciation of the exchange rate has not been the main mechanism of adjustment. The more common adjustment option has been to maintain a relatively fixed exchange rate and increase total absorption or expenditure. The government, which is the main recipient of the foreign exchange, can either import directly or spend the money domestically. Domestic public spending financed by oil revenues can be regarded as domestic deficit expenditure since it is over and above the domestic revenues of the government derived from taxation. The idea of domestic deficit expenditure is based on the concept of the domestic budget balance, a concept which is sufficiently central to the empirical chapters which follow as to require a more detailed discussion.[6]

In countries in which a substantial volume of external receipts and payments passes directly through the government's budget, the domestic budget balance provides a better approximation of the impact of budgetary expenditures on money supply than the overall budget balance. The separation of government transactions into foreign and domestic transactions represents an attempt to estimate the direct impact of the budget on domestic, rather than total, demand.

Government expenditure abroad does not add directly to domestic demand and therefore does not affect domestic employment and output. Similarly, government receipts from abroad do not directly reduce private domestic resources.

The concept of domestic budget balance, which emphasises the first-round effects of budgetary operations on the generation of income and purchasing power, has direct implications for the analysis of monetary policy. It can be demonstrated that the impact of the government budget on money creation is determined by the domestic budget balance. The overall budget balance B is the difference between total expenditures (domestic G_d plus foreign G_f), and total revenue (domestic R_d plus foreign R_f).

Symbolically:

$$B = G_f + G_d - R_f - R_d \tag{28}$$

or

$$B = (G_d - R_d) + (G_f - R_f) \tag{29}$$

where $(G_d - R_d)$ is the government's domestic budget balance and $(G_f - R_f)$ is the government's foreign balance. The overall balance is financed through foreign borrowing, ∂F_g where the delta prefix stands for absolute change in the variable during the accounting period); domestic non-bank borrowing ∂L_d; and net credit from the domestic banking system ∂C_g. Assuming that domestic non-bank borrowing is unimportant, then:

$$(G_d - R_d) + (G_f - R_f) = \partial F_g + \partial C_g \tag{30}$$

or

$$(G_d - R_d) = \partial F_g - (G_f - R_f) + \partial C_g \tag{31}$$

The government's foreign balance and foreign borrowing will represent the change in net foreign assets attributable to government's operations, ∂NFA_g. Thus:

$$(G_d - R_d) = \partial NFA_g + \partial C_g \tag{32}$$

Since ∂NFA_g and ∂C_g are components of domestic liquidity, the direct impact of the government budget on liquidity creation is determined by the domestic budget balance. Any non-bank domestic borrowing by the government would enter negatively on the left-hand side of Equation (32).

Now consider the conventional framework for tracing the process of money creation:

$$\partial M = \partial C_p + \partial C_g + \partial NFA + \partial NUA \tag{33}$$

where M is domestic liquidity; C_p is the net claims of the banking system on the private sector; C_g is the net claims of the banking system on the government; NFA is the net foreign assets of the banking system; and, NUA is the net unclassified assets of the banking system. To arrive at an alternative presentation suitable for highlighting the role of the domestic budget balance, ∂NFA needs to be divided into the change in net foreign assets attributable to private sector operations ∂NFA_p, and to government operations ∂NFA_g. Introducing this distinction, and substituting Equation (32) into Equation (33), we have:

$$\partial M = \partial C_p + \partial NFA_p + (G_d - R_d) + \partial NUA \tag{34}$$

where ∂M is the change in domestic liquidity (money plus quasi-money); ∂C_p is the change in the claims of the banking system against the private sector; ∂NFA_p is the balance of payments of the private sector; $(G_d - R_d)$ is the domestic budget balance; and ∂NUA is the change in the net unclassified assets of the banking system. Equation (34) states that the main determinants of the money supply are the domestic budget balance (deficit or surplus), the balance of payments of the private sector (deficit or surplus), and the changes in domestic bank credit to the private sector. Also note that government's external transactions do not appear in the identity, reflecting the fact that they do not contribute directly to the money-creation process. The rise in the level of the domestic deficit which increases the money base and the consequent operation of the money multiplier would result in an increase in the money supply minus the leakages via external transactions of the private sector. It should be added that government borrowing from the banking sector would be accounted for in the calculation of the domestic deficit. The final element of the money supply is the net unclassified assets of the banking system, often small in comparison to the other elements which have been emphasised.

To clarify the above symbolism, consider the following hypothetical example. Briefly, changes in the money supply are essentially dependent upon two main variables, as presented in the following hypothetical table:

Factors affecting changes in the money supply

		Y_t	Y_t+1
1	Domestic budget deficit	100	200
2	Private sector's impact		
	(a) external transactions	-40	-80
	(b) bank credit to the private sector	50	90
	Changes in the money supply	110	210

Such domestic deficit expenditure results, therefore, in a wave of monetary expansion which will have inflationary consequences – raising the level of domestic demand via its own direct effect as well as through the operation of the multiplier. The rise in aggregate demand plus the inelastic supply of non-traded goods would bring about a real appreciation, as discussed earlier.

As has already been emphasised, it is not only the aggregate level of government expenditure but also its pattern which is of considerable importance. In other words, it is not only how much the government spends but how the government spends this money and more importantly, which sectors benefit from this expenditure and what in turn is their expenditure pattern. This point brings us to the implications of the specific policies, programmes, and projects of the government and their development strategy with fundamental implications for the type of economic adjustment and the consequent economic structure and supply response during and after the oil boom.

Sectoral allocation of investment

The intra-sectoral as well as the inter-sectoral pattern of public expenditure can critically affect the output of the various sectors of the economy. In the particular case of oil economy of a less-developed country (LDC), with the direct accrual of oil revenues to the government, the government can now undertake its own usual types of investment which is usually in infrastructure at an expanded scale. The government, therefore, tends to invest in large infrastructural projects and public capital expenditures tend to become largely concentrated in the non-traded goods sector. Also the government is now increasingly able to undertake military expenditures.

The increase in infrastructural investment is further reinforced by two basic forces: first, with the foreign exchange constraint relaxed, the non-traded goods sector becomes the main constraint, and second infrastructural projects are large and require substantial amounts of resources and, hence, are an easier way of spending large sums of money.

Infrastructural projects become of paramount importance especially because the traded goods sector can now be substituted through imports. This is an important point because it is not only private investments which move towards the non-traded sector, but also there is a substantial increase of public investment in that direction. Indeed a purely economic explanation of the adjustment process, as already discussed, which relies on relative price changes to explain resource allocation, would miss the bureaucratic allocation of resources to the non-traded goods sector by the state.

This form of resource allocation by the state to the non-traded goods sector is perhaps as important as, if not more important than, the shift of private investments towards the non-traded goods sector. Such high levels of expenditure in the non-traded sector by both the public and the private sector can only be justified if the high oil-revenue levels continue for a long time and traded goods are easily available through domestic production or through imports. This, however, is unlikely to be the case. The domestic supply of traded goods is likely to decline as a result of the real appreciation of the exchange rate and the direct shift of resources away from that sector via government policy. Also, with the slowdown and halt of oil revenues due to price fall or depletion, imports are likely to decline. If the economy is not prepared to deal with the decline of oil revenues through some other source of foreign exchange or domestic production of traded goods, the country would be faced with a massive non-traded sector which is poorly maintained and largely inappropriate for the post-oil economy due to the shortage of foreign exchange. The overemphasis on infrastructure, and especially the massive type, is one of the 'pitfalls' which the government could avoid with an alternative set of policies.

Apart from the inter-sectoral allocation of resources between the traded and non-traded goods sector, there is also the question of the intrasectoral allocation of resources within both the traded and the non-traded goods sector. A basic factor in this context is the development strategy which is followed by the government.

Development strategies can be broadly characterised as either 'unimodal' or 'bimodal'. Unimodalism involves a progressive transformation of the entire economy, sector by sector, through the gradual evolution of institutions and technologies compatible with uniform development of each sector. Unimodalism also involves using the public sector more as a co-ordinating body than as an executing body. This means that a unimodal strategy puts greater emphasis on broad private sector participation and market development while the public sector becomes complementary to the private sector.

Bimodalism, on the other hand, involves the concentration of resources in large-scale capital intensive sub-sectors of the economy, usually under the umbrella of direct government support for crash modernisation projects with a skewed pattern of investment as the result. Under this strategy government is likely to become directly involved in production. In this sense, bidmodalism relies on the government more as an executing body than a co-ordinating body. Moreover, the large-scale and often capital-intensive nature of this strategy involves not the evolution but mostly the transfer of institutions and technologies from abroad. In practice, however, bidmodalism usually results in the parallel growth of a large-scale capital-intensive sector along with the growth of an informal small-scale sector with spontaneous and autonomous orientation, in spite of the neglect and even positive discouragement of the state to the contrary. Indeed the failings of the large-scale formal sector allow considerable space for the spread of the informal sector.

The development strategy which is followed in an oil-exporting country, especially in the boom phase, is likely to be largely bimodal in practice and this can critically alter the supply response of the traded sector in particular. In agriculture, for example, the choice of large-scale mechanised farming by the state or joint venture with foreign capital is likely to prevail over support for small farmers. This is because the state is likely to see large-scale projects as a faster way to increased production, providing better central control, and having fewer institutional problems such as organising an effective credit, marketing, and extension network to help the small farmer. Larger projects, as mentioned above, also allow for faster lump-sum disbursements than small projects.

Bimodalism in agriculture, however, can result in crowding out the small-scale peasant farmer in terms of essential inputs such as fertilizer, research and extension, water, credit, storage and transport. This can lead to a considerable loss of both existing and potential agricultural output. Bimodalism in agriculture also has other ramifications, especially for the provision of non-traded inputs to agriculture. For example, the concentration of large dams as opposed to assistance to small-scale irrigation, or building modern tarmac roads for inter-urban transport as opposed to rehabilitation and reconstruction of essential rural feeder roads, puts the small farms in a position of further disadvantage with implications for the supply response of that sector. Again this is a potential policy 'pitfall' which arises since capital is temporarily cheap and the foreign exchange constraint temporarily relaxed,[7] which can be avoided.

Land-ownership policy

Land is a non-traded factor of production. The price of land, especially urban and peri-urban land would rise with adjustment to higher oil revenues because of derived demand for land due to the increase in urban construction and peri-urban non-traded agricultural production such as perishable fresh vegetables. Rising land values would raise the cost of production for land-intensive activities. Moreover, the government, by becoming a major investor due to its vastly increased revenues, also becomes a major claimant of land for its own purposes and increasingly uses legislative power to appropriate land. In other words, although the value of land rises because land is a non-traded good, the actual commercial value would depend upon government policy regarding land-ownership. The availability of land titles and security of land-ownership affect the growth of the land market and therefore the level of investment at the farm level. Government decrees concerning the nationalisation of land can lead to arbitrary seizure of land by the government, hence resulting in lowering the confidence of investors at the farm level.

Credit policy

A critical factor in supply response is access to cost-reducing technology as well as working capital, which in turn depends upon access to credit. Government's credit policy can play a key role in affecting the fortunes of industries by allowing or depriving them of access to credit. Credit policy is also significant for small farmers who are the main food producers This is because informal credit is usually difficult to obtain, quite expensive, and leaves little for the farmer for reinvestment. Also, informal credit usually meets only the working-capital requirements of the farmer, leaving a gap as far as the credit for the adoption of new technologies is concerned.

During the oil boom, bimodalism is likely to affect credit policy, hence the bulk of official credit for the traded goods sector is likely to end up in large-scale capital-intensive sub-sectors of industry and agriculture. Although capital-intensive investment may be considered as the correct response in the short run, it is unlikely to be the correct response in the medium and long term (the medium term being the period in which the country experiences sharp changes in the terms of trade for oil and the long term being the period after which oil is depleted). Both because of the highly fluctuating nature of oil revenues in the medium term and the possibility of depletion in the long term, dependence on capital-intensive and import-dependent industrialisation is likely to result in difficulties for economic management. With a drop in the oil price, the government could experience foreign exchange difficulties and be unable to import the required raw materials and spare parts for the operation of such industries. In fact, given the temporary nature of oil revenues, it is all the more important that at least the government adopts a perspective on investment which accommodates the possibility of sharp falls in oil revenues.

Moreover, commercial non-government bank lending is likely to follow the relative price signals and shifts towards private non-traded activities such as import financing, construction and transport. Bimodalism in credit policy and the consequent skewed pattern of credit distribution would therefore reinforce the adverse consequences of bimodalism already described. Again this is a policy pitfall which could be avoided.

Pricing policy for food crops

With vastly expanded financial resources, the government can effectively implement an agricultural pricing-policy programme. The government can set prices because it can both engage in substantial purchases in the domestic market and resort to imports in order to maintain the minimum price. The higher or lower official price can therefore become an important signal for changes in private sector production incentives. The effectiveness of agricultural pricing policy would also depend, however, on whether the agricultural commodities in question are traded or non-traded. If the main food staple is wheat, for example, the possibility of controlling its price is much greater than if the main food is cassava or millet, which are not commonly traded internationally due to high bulk and low value and their price can only be controlled indirectly through the import of substitutes such as wheat and rice into the country. In the case of an oil-exporting country, the government can exert considerable depressing influence over the price of traded crops by direct imports. This is because the government, as in many other countries, could in this way lower the cost of its own food subsidies. The main point here is to emphasise that the cheapening of traded goods through the real appreciation of the exchange rate should be analysed in conjunction with government's policy regarding food imports and the traded or non-traded character of food crops.

Tariff policy and import control

With higher domestic cost due to the higher price of non-traded goods, at world prices, the domestic exports and import-competing goods are no longer competitive. The higher comparative cost of production in the oil-exporting country gives an advantage to competing countries' exports to the oil economy. The government in the oil-exporting country, however, can grant tariff or limit imports on a selective basis for the protection of traded goods.[8]

The tariff or the import control would increase the range of prices at which a commodity is non-traded and, therefore, sheltered from international trade. This point is illustrated in Figure 1.8, based upon Figure 1.1 which was used earlier to define non-traded goods.

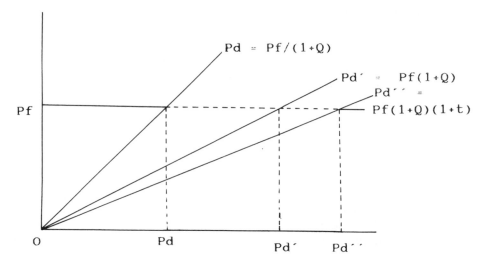

Figure 1.8

An import tariff in the home country at rate t would widen the range of prices at which a commodity is non-traded. For the commodity to be imported now, the foreign price inclusive of transport charges and tariffs is $P_f(1 + Q)(1 + t)$ which would have to be less than or equal to the domestic price. This defines the maximum price at which the good is non-traded before it becomes imported, $P_d = P_f(1+Q)(1+t)$. The cone that defines the region of non-traded goods now has a larger angle since it includes a larger range of prices. The analysis for quantitative import control is similar since such measures also raise the range at which a commodity is non-traded.

In the oil boom, the government is likely to reduce the average level of tariffs and generally adopt a more liberal import policy given its favourable balance of payments. This can be regarded as a means of 'adjusting' to the balance-of-payments surplus which involves reducing the levels of protection. The basis of this argument is the idea that the decline in the traded sector should be concentrated in areas with least comparative advantage – areas with highest protection. In contrast to exchange-rate adjustment which affects the prices of traded goods as a whole, protection only raises the prices of selected traded goods. Reduction of tariffs results in a switching of expenditure towards imports and, hence, reducing the financial surplus in the balance of payments.

Reduction of tariffs in a less developed economy faced with real appreciation is a controversial point. The opponents of lower tariffs could argue that, if anything, levels of protection should be increased so as to shield the traded sector against the adverse impact of real appreciation. Furthermore, the absence of an active capital market in the LDCs plus significant gaps in information precludes the possibility of complete reliance on the market forces for the build-up of a non-oil traded sector large enough and in time to provide the economy with sufficient foreign exchange in the absence of oil. This is in essence an infant-industry argument which is further strengthened because of the macroeconomic context of an oil economy.[9] Protection may further be justified by pointing out that the problem is not only one of long-term depletion but also insuring against short-term fluctuation in the price of oil. The role of the government in determing the tariff and import control policy is evidently critical for the outcome of 'adjustment' in the output of the traded goods sector.

Consumer subsidies and other current expenditures

Consumer subsidies, usually considered as 'current' expenditure by the government, can prevent the relative price rise of specific commodities and, depending upon the importance of that item sectorally, consumer price subsidies can lower the sectoral price level. Consumers would, therefore, divert the extra funds thus released to the purchase of other non-subsidised goods which could lead to an additional rise in the aggregate level of prices. Nevertheless, if the subsidised commodities form a major part of total consumer expenditure, the subsidies could cause the price level to be lower than otherwise and thereby lower the pace of real appreciation. Also, the change in the consumption habits of the population brought about by the cheaper subsidised traded consumption goods such as food items, plus the shift of demand away from traditional foods, has fundamental implications for the sustainability of such demand, as well as the production structure of traditional agriculture. These possibilities again demonstrate the importance of discussing the composition of government expenditure and its consequences for the structure of prices and output in an oil economy.

Another type of current expenditure by the government is the purchase of such traded goods which are not consumer goods. An example is military expenditure by the government. In this case consumer welfare is hardly affected, hence no change in the relative price of traded goods is brought about unless additional domestic military expenditure follows the military purchases from abroad. Nevertheless, from a supply-side point of view, military expenditure crowds out resources from productive expenditure and thereby affects the cost of production for non-military goods and services.

This concludes our theoretical discussion of the underlying political economy of adjustment to higher oil prices. The following chapters are an empirical investigation of the theoretical issues raised in this chapter, with the purpose of evaluating the consequences of the policies followed for the productive structure of the economy, keeping in mind the need to transform the oil wealth into a renewable resource base.

Notes

1. For a recent theoretical survey of the subject, see Corden (1982). For further conceptual background, see Mackinnon (1976). For a more formal elaboration of this theme, see Corden and Neary (1982). For an excellent theoretical exposition and an important source for this chapter see Dornbusch (1980, Ch. 6). For a short exposition of the Dutch-Disease model with emphasis on income distributional aspects, see Caves and Jones (1983, Ch. 6).
2. The concept of specific capital in the short run is from the so-called 'Specific Factors Model'. This is a model based on the assumption that not all factors can move freely between industries as incentives for reallocation arise. Some factors are temporarily locked in and cannot be adjusted immediately. Only over a long period of time, if incentives for reallocation continue, will the initially locked-in factors become mobile, and a gradual adjustment to equilibrium take place. This also gives rise to the possibility of short-term persistence of differential returns to capital in different sectors of the economy. For a formal exposition of the specific factors model see Mayer (1974).
3. Proof:

Equation (1) states:

$$L_t(W/P_t, K_t) + L_n(W/P_n, K_n) = L$$

differentiating Equation 1 gives:

$$L_t'[(P_t dW - W_d P_t)/P_t] + L_n' [(P_d W - W_d P_n)/P_n] = 0$$

rewriting in terms of percentage changes $W\sim = dW/W$ and $P\sim = dP/P$:

$$L_t' (W\sim W/P_t - P_n\sim W/P_t) + L_n' (W\sim W/P_n - P_n\sim W/P_n) = 0$$

or

$$(W{\sim} - P_t{\sim})L_t'W/P_t + (W{\sim} - P_n{\sim})L_n'W/P_n = 0$$

Defining E_i as the elasticity of labour demand or $E_i = (L_i'/L_i)(W/P_i)$

and defining S_i as the share of the labour force in sector i or $S_i = L_i/L$

we have:

$$(W{\sim} - P_t{\sim})LS_tE_t + (W{\sim} - P_n{\sim})LS_nE_n = 0$$

or

$$W{\sim}(S_tE_t + S_nE_n) = P_t{\sim}S_tE_t + P_n{\sim}S_nE_n$$

If a new parameter z is defined as $z = E_nS_n/(E_nS_n + E_tS_t)$ then

$$W{\sim} = ZP_n{\sim} + (1 - Z)P_t{\sim}$$

4. Workers fired from the traded goods sector would have to be absorbed in the non-traded sector. Of course, the extent of labour absorption in the non-traded goods sector would depend on the extent of fall in wages relative to the price of non-traded goods. For example, if wages are indexed to the Consumer Price Index, (CPI), the extent of the real wage fall in the non-traded sector would depend upon the composition of the consumer expenditure in terms of traded and non-traded goods. If traded goods make up the bulk of consumer goods, given the indexation of wages to the CPI, then an increase in real wages in terms of traded goods will allow a large decrease in the real wage in terms of non-traded goods. This would allow the non-traded sector to more than absorb the workers fired from the traded goods sector. However, if the share of traded goods is large in the total consumption basket of the workers, wage indexation would mean a higher real-wage level in terms of non-traded goods. This means that with an increase in W/P_t, a high share of non-traded goods in total consumer expenditure, and wage indexation to the CPI, less workers would be absorbed in the non-traded sector and unemployment can result. For a detailed formal discussion of this point see Van Wijnbergen (1984).
5. The 'magnification effect' comes from Jones (1965, pp. 561-2). For a formal analysis of magnification effects in the specific factors model, see Evans, (1986, Appendices 4.4 and 5.3). A simple and concise discussion and an important source for this section is Caves and Jones (1983, Ch. 6).
6. The following discussion is based on Morgan (1979).
7. Note that this argument is intended for the larger oil economies with a shorter depletion horizon rather than for the smaller ones with massive reserves which will last well into the next century.
8. If protection were raised sufficiently to ensure that the protected sector was not harmed by the petroleum boom, even though the real exchange rate had appreciated, the burden of adjustment would then fall entirely on the non-protected traded sectors such as small-holder and export-crop agriculture, which would be harmed by more than they would be if full adjustment had been brought about with a constant level of protection. The real appreciation required for adjustment to higher oil revenues would then be greater than if protection levels had stayed constant. Tariffs on manufactures would therefore tend to produce further appreciation..
9. An excellent discussion of the infant industry argument may be found in Corden (1974, Ch. 9).

PART II
THE CASE OF IRAN

2 Adjustment policy, development strategy and macroeconomic impact on Iran in the 1970s

In this chapter the adjustment policy, the underlying development strategy, and the consequent macroeconomic impact on the non-oil Gross Domestic Product (GDP) in Iran are described and analysed. The chapter consists of three main sections. In the first, the process through which a higher aggregate level of expenditure is brought about is explored. The transformation of the foreign exchange earnings into domestic money through the increase in domestic budget deficit, credit expansion by the banking system, and the rise in the level of aggregate expenditure are discussed.

The second section turns to the development strategy and the composition of expenditures undertaken by the government. It will be shown that high levels of public expenditure reflected an ill-conceived, big-push strategy of development, poor planning and budgeting, a personalised top-down approach to decision-making, and the abandonment of detailed project-appraisal procedures for project selection. The point here is not that such an approach is specific to the oil boom but rather that easy money facilitated such an approach which in turn affected output and resource allocation in the economy adversely, again back to the point that both the form and the content of government actions are critical for determining the adjustment process.

In the third section, the GDP, and particularly its non-oil component, is brought under detailed analysis. The reason why the GDP and not the GNP is analysed is that the main focus of this book is on what happens to domestic production and not to the total income of domestic residents. In other words, net factor payments from abroad do not feature significantly in the analysis of production in the oil-exporting country, although some discussion of capital payments by the government as an option for dealing with the oil revenues is included in the analysis. The purpose of the third section is to determine the extent of the differential growth rate of traded and non-traded sectors. Relative price changes in the different sectors of the economy are analysed in order to determine the extent of real appreciation or the relative price increase of non-traded goods, resulting from the rapid increase in domestic demand. Moreover, the growth of real output of the different sectors is analysed in order to assess the supply responsiveness to relative price changes.

Oil revenues and changes in aggregate expenditure

As can be seen from Table A.1, by 1969 Iran was already highly dependent on oil exports, which accounted for 72 per cent of her total export earnings; on average, from 1969 to 1972 the share of oil and gas exports in total exports was 75 per cent. The rapid rise in oil and gas exports begin during the year 1973–4, when revenues double to $5.160 billion, continuing in 1974–5, when they reach $19.235 billion, which raises their share of total exports to 91 per cent. The highest level of export revenues from oil and gas was obtained in 1977, amounting to $21.096 billion for 2 billion barrels of oil; the average share of oil and gas exports in total exports between 1973 and 1978 was 84 per cent. From 1979 to 1981, oil exports declined to about 500 million barrels per year, picking up in 1982 to 1 billion barrels or half the volume of exports of the mid-1970s.

The share of oil in GDP at current market prices changed from 17.2 per cent in 1969 to 53 per cent in 1974 and declined to 28 per cent and 17 per cent of the GDP in 1979 and 1982 respectively.[1]

In Iran, the oil industry is nationalised and the government is the sole recipient of the foreign exchange receipts from oil sales. With the rise in oil exports, government revenues in current prices also increased from 172.3 billion rials in 1970 and 464.8 billion rials in 1973 to a high point of 2,034.2 billion rials in 1977 after which they declined along with the fall in oil revenues and picked up again with the rise in the revenue in 1982 (see Table A.2). Using the consumer price index as the deflator and 1974 as the base year, the real increase in government revenue is from 242.6 billion rials in 1970 to 1,394.4 billion rials in 1974, and then to 1,640 billion rials in 1976; after 1976, although government revenue increases in nominal terms, it declines in real terms to 560 billion rials in 1981, and to 882 billion rials in 1982, in spite of the pick-up in oil sales in that year (Table A.2). Using any other deflator such as the GDP deflator does not significantly change the main trend in government real revenues.

The rapid rise in government revenue at the initial phase of the boom is linked to the government's central role as the main recipient of the oil revenues. It is therefore essential to examine the utilization of this revenue which determines the form and consequences of adjustment. In Iran, the government chooses to spend most of the revenues domestically while maintaining a relatively stable nominal exchange rate to the dollar. There was, of course, some capital outflow, in the form of sterilisation of the export earnings (non-adjustment) through investment abroad, but this was not the main policy option. Taking capital payments (capital movements including monetary movements) as a percentage of total receipts (current plus capital), Table 2.1 shows that, between 1974 and 1982, 21.5 per cent of total foreign exchange receipts were transferred abroad; therefore more than 78.5 per cent were utilised in other ways.

Table 2.1 Capital flows abroad ($ billions of dollars)

	1974	1975	1976	1977	1978	1979	1980	1981	1982	1974–82
Total receipts	21.6	22.9	26.0	29.2	25.0	23.8	14.8	15.4	21.4	200.1
Capital payments	5.8	4.5	2.6	2.1	3.9	1.3	9.5	8.0	5.4	43.1
Capital payments as % of receipts	26.8	19.6	10.0	7.2	15.6	5.5	6.4	52.0	25.6	21.5

Source: Bank Markazi Iran, *Annual Report,* 1972, 1976, 1979, 1983

Since 'non-adjustment' or the sterilisation of the reserves was not the main option chosen by the government, other options can now be considered. Of the three options discussed in Chapter 1, namely, explicit exchange-rate appreciation, higher domestic absorption, and lower tariffs, the raising of the level of domestic absorption appears to have been the most important accompanied by lowering of selected nominal tariffs.

Consider the nominal exchange rate (the real exchange rate is discussed at the end of this section). In 1973, the currency appreciated from 75.75 rials to the US dollar, to 68.88 rials to the dollar or by about 10 per cent. In 1976, the rial depreciated from 67.63 rials to the dollar to 70.22 rials to the dollar or by 3.7 per cent, and the trend, until 1980, was generally a slight fall relative to the dollar.[2] Considering the large increase in Iran's foreign exchange earnings during this period, the nominal fluctuation of the rial has been rather marginal and, therefore, it cannot account for adjustment. Moreover, until 1978, there was hardly any black market for the rial; from 1975 until 1978, the government liberalised capital movements and so ample foreign exchange was available at the above rates. Since 1978, however, the reintroduction of exchange controls has significantly raised black-market rates.

Having shown that neither sterilisation nor explicit revaluation have been utilised to any significant extent as a response to the rise in financial surpluses, the other remaining options are the lowering of tariff barriers and/or increased domestic expenditure. There was a general lowering of tariff barriers after 1974 in Iran and this point is discussed in more detail in Chapter 3. The lowering of tariff barriers, however, was used in conjunction with increased domestic expenditure. Let us consider the question from a monetary angle, where the government's domestic deficit initiates the process of liquidity expansion, thereby increasing the level of expenditure.

It can be seen from Table A.3 that, in all the years between 1970 and 1980 except 1974, total government expenditure was higher than total government revenue. Here the concern is not with overall deficit but with the *domestic* deficit, which is the difference between government's domestic revenues and its domestic expenditure. After calculating government's domestic deficit (see also Table A.3) from the budgetary and the balance-of-payments account, a test was conducted to determine the importance of the government deficit in the determination of the total broad money supply (M2). The other component of the money supply is the private sector's net liquidity contribution which is the net result of the expansionary effect of bank credit to the private sector and the contractionary effect of the private sector's balance-of-payments deficit. It may be added that, contrary to Jakubiak and Dajani (1976), we found that bank credit to the private sector has been greater than the private sector's balance-of-payments deficit and therefore bank credit to the private sector has had an expansionary effect on liquidity in the economy (Table A.4).[3]

The government's domestic budget deficit grew at 40 per cent per year during the 1970s. Its absolute size varied around 20 per cent of GDP, and it grew from being almost one-third of the total export earnings to about two thirds in 1975–76 and even surpassing total export earnings in 1979–80. These comparisons are only illustrative of the massive size of the domestic budget deficit.

Comparing the periods 1962-69 and 1970–80, the impact of the level of the domestic budget deficit on the change of money supply becomes quite clear. From 1962 to 1969, the domestic budget deficit rose from 16.7 billion rials to 45.5 billion rials a level less than three times as high. From 1970 to 1980, however, the domestic budget deficit grew from 38.3 billion rials to 1,363.1 billion rials, which is more than multiplying the a 35-fold rise (Table A.3). Prices in the same period rose by about four times (Table A.7). The sharp jump begins markedly in 1973. From 1973 to 1977 the domestic budget deficit grows by about ten times and the money supply grows by about four times. The net liquidity contribution of the private sector in the same period is calculated to have risen by about three times. Quantifying the relative weights of the domestic budget deficit and the liquidity impact of the private sector in the equation below we find that for the 1962–80 data:

ln(M2) = 0.65 + 0.25 (ln*DBD*) + 0.75(ln*PSC*)

        (5.0)   (5.9)   (14.0)

where *DBD* is the domestic budget deficit (Table A.3) and *PSC* is the private sector liquidity contribution (Table A.4);

R^2 = 0.99, and the Durbin–Watson statistic is 1.44 (adjusted for auto-correlation, Cochrane–Orcutt method).

    The domestic budget deficit (*DBD*) appears to have less relative impact on changes in the money supply (M2) than the relative impact of the net liquidity contribution of the private sector. The private sector emerges as having the greater relative influence on the money supply.[4] Again this result is different from the findings of the study by Jakubiak and Dajani (1976), who claim that the government's net domestic deficit is the 'main factor accounting for the acceleration in the rate of monetary expansion' (p.13). What is more plausible however, is that the government is the source of initial growth (through rial spending of its dollar deposits at the Central Bank) of the money supply but the liquidity impact of the private sector's demand for money should not be underestimated.

    Having discussed the key determinants of the money supply, let us now consider the growth of domestic expenditure and the consequent changes in the price level and in relative prices. The implicit assumption behind linking the money supply and the growth of expenditure is that money demand grows more slowly than the money supply. In the more extreme form of monetarist approach, the assumption is that there exists a stable demand for money which is a function of income. If the money supply grows faster than the money demand, as implied by the growth of income, then the excess money balances would be spent on goods, services and financial assets, which in turn would drive up their prices. All that is necessary for our purposes, however, is that money demand does not grow as fast as the supply of money and although the evidence may not indicate a functionally stable demand for money, it is nevertheless consistent with the money supply growing faster than money demand.[5]

The development strategy and the composition of expenditure

The rapid growth of expenditure after the oil boom was largely justified through a big push strategy of development. High growth and relative price stability in the 1960s and the success of the Fourth Plan had combined to allow a highly optimistic attitude towards planned targets. The Shah, with his aim of a 'Great Civilisation' and a return to the greatness of Persia, announced that the dream was now a reality and that there would be a massive effort on all fronts so that it would be achieved. Indeed, the speed with which the big push was to be implemented was to be of paramount importance in the shaping of subsequent expenditure patterns.

In early 1974, Iran had almost completed the first year of its Fifth Development Plan which was to run from 1973 to 1978. The sudden upsurge of oil revenues, however, necessitated a revision of the Plan. The planning office was therefore instructed to prepare a major revision in six months. (Razavi and Vakil 1984, Ch. 4).

The already prepared Fifth Development Plan was in itself a rather ambitious document calling for a targeted real growth rate of 11.4 per cent per annum. Table 2.2 summarises the Plan and the revision.

Table 2.2 Fixed investment by sector, 1973–7, original and revised (billions of rials)

Sector	Original	Share(%)	Revised	Share(%)	Revised as % of original
Industry & mines	552.2	22.2	845.8	18.0	153.2
Agriculture	180.2	7.2	309.2	6.6	171.5
Transport	188.3	7.6	492.1	10.5	261.3
Housing	402.3	16.2	924.8	19.7	229.9
Oil & gas	461.0	18.5	791.1	16.8	171.6
Others	702.7	28.3	1,334.5	28.5	189.9
Total	2,486.7	100.0	4,697.5	100.0	189.9

Source: Plan and Budget Organisation (1975) $1 = 67.5 rials

In fact the original plan was already close to absorptive capacity (Razai and Vakil 1984).[6] The problem facing the planners, therefore, was how to determine the maximum possible level of investment which could be productively absorbed by the economy while maintaining the relatively smooth growth rate of the previous decade (see Table A.10).

The Plan and Budget Organisation (PBO) provided a number of scenarios for achieving a macrobalance. Interestingly, within the PBO there were opposite forces at work; since the Planning Division and the division in charge of projects were in disagreement as to the best course to pursue. The former advised against a maximum-expenditure approach, while the latter pursued a maximalist line, only worrying about the allocations between sectors. Three scenarios presented were based on a 31 per cent, 98 per cent, and a 141 per cent increase in public investment. The requests from the ministries even exceeded the high-expenditure scenario by 945.0 billion rials ($14.0 billion) (Razavi and Vakil 1984).

The final meeting which approved the maximalist strategy took place in Ramsar in mid-1974 and was attended by the Shah. Against the advice of many economists concerning the limits of the country's infrastructure and manpower, that it, absorptive capacity, a crash programme was to be implemented immediately. The economists' objection that such a rapid increase in expenditure would create severe shortages, in manpower and infrastructure was apparently dismissed with contempt by the Shah. His answer was entirely based on the idea that everything can be imported if necessary. At the end of the meeting he told the audience: 'The 'Great Civilisation' we promised you is not a utopia. We will reach it sooner than we thought. We said we will reach the gates in 12 years; but in some fields, we have already crossed its frontiers'. (Graham 1979, Ch. 5).

As can be seen from Table 2.2, the revised version resulted in 89 per cent increase in planned expenditure. Moreover, looking at the distribution of this expenditure in the different sectors of the economy, it is clear that the shares of agriculture, industry, oil and gas declined while the housing and transport sectors had an increased share. Housing and transport were seen as bottleneck areas by the planners, and rightly so.

What is significant from the point of view of this book, however, is that there was an arbitrary increase in public expenditure without sufficient appraisal of its feasibility and impact, and that housing and transport, which are both non-traded sectors, increase their share in total expenditure. This move shows how with the boom public investment also shifts in the direction of non-traded goods in order to alleviate the shortage which was created in the first place with an inappropriately high level of expenditure.

Moreover, government consumption expenditures rose sharply. Public consumption expenditures rose from 427.9 billion rials ($6.3 billion) in 1973 to 628.3 billion rials ($ 9.2 billion) in 1974 or 540 billion rials at 1973 prices (Table A.6). This means that from the total government budget for 1974 (1,254 billion rials), 50 per cent went into current or consumption expenditures, with military expenditures accounting for over half of it and subsidies accounting for a major part of the rest.[7]

The most important subsidies were for maintaining a low consumer price for wheat and sugar. For example, subsidies on imported wheat amounted to about 50, 53 and 57 per cent of CIF prices in 1974, 1975, and 1976 respectively (Shafaeddin 1980, pp. 149–50).

It should be added that this approach to the use of oil revenues, putting aside military expenditures, contained a fundamental error concerning the nature of the scarce factor in the economy. The oil boom removed one constraint, namely foreign exchange. It is generally true, however, that if one constraint is suddenly removed, one simply encounters the next one, be it domestic saving, skilled and unskilled manpower, management, transport capacity, or whatever. This is not to say that there is no substitutability amount the scarce factors but that proper substitution requires considerable time and planning, a factor ignored by the government at this time.

The original idea of the big push as conceived by authors such as R. Nurkse (1964) was a means of overcoming the vicious circle of poverty: a country is poor because savings are low, hence investment is low, hence there is no growth, hence poverty continues. In other words, the scarce factor was viewed to be finance or savings. In the case of Iran, clearly the problem was not to generate savings but how to utilize the existing financial surpluses so as to acquire macrobalance and diversification. Moreover, the Shah's view that everything can be imported implies that foreign savings can also finance non-traded goods. In the short run, however, this is an erroneous view. The error arises from conceptualising the economy through a one-sector model without distinguishing between traded and non-traded commodities. In the short run, it is important to distinguish between traded and non-traded investible resources, since although, in terms of national accounts, investment equals savings plus the trade balance, substitution between traded and non-traded goods requires that the price of a good passes through the non-traded phase of Figure 1.1 (in Chapter 1) before it becomes traded or in this case imported.

The fact that there is a limit to the rate of growth of investment which is set by all domestic factors of production which are non-traded since they face rapidly rising import cost problems was not recognised. Rapidly rising costs implies rapidly falling returns on investment involving bottlenecks in infrastructure, public services, and waste of resources.

Having discussed the plan during this period, let us now consider the government's budget. Normally the budget document consisted of an annual economic report for which the Planning Division had the primary responsibility, with co-operation of the Central Bank and the Ministry of Finance. The final figures for the 1974 budget, however, were prepared in a series of hurriedly conducted meetings without the consultation of the Planning Division and with total disregard for the Fifth Plan (Razavi and Vakil 1984, p. 79). A glance at the 1974 budget would show the absence of economic rationality in that document. The planned expansion called for a 140 per cent increase while the actual expansion amounted to a 162 per cent increase over government expenditure in 1973 (Table 1.3). Thus began a two-year economic boom in Iran.

The spending spree of various industries and the defence establishment knew no bounds in this period. The need to spend the allocations meant that sub-optimal projects were quickly approved. In fact, in this period the power of the Co-ordination and Supervision Division Committee which was in charge of maintaining standards for project appraisal and implementation was greatly reduced and the Division found itself exerting very little control over the budget. For the main projects which absorbed the major part of the government budget, the concerned agency would first obtain the royal decree and then send the proposal down to the Co-ordination and Supervision Division largely as a *fait accompli*. The spending spree therefore resulted in a virtual abandonment of the detailed appraisal methodology and the committee in charge was forced to approve large projects mostly dictated from above (Razavi and Vakil 1984, p. 45). The Co-ordination and Supervision Committee was thus left with a residual budget to meet the hundreds of requests from below for development projects in the essential sectors of the economy. The final outcome was massive allocations to large industrial, infrastructural and luxury projects, many of which were only suited for satisfying the Shah's obsession with symbols of progress.[8]

It is important to note that the components of public expenditure are an essential factor in affecting the outcome of the adjustment process. These components are usually a reflection of the development strategy pursued and they have a significant impact on the supply response of the economy. A third of the planned expenditure in the Fifth Development Plan, as was shown in Table 2.2, was allocated to housing and transport. In other words, an increase in government expenditure meant an increase in investment for producing non-traded goods, at least in the first round, such as infrastructure and services. Although the details of the underlying projects are not discussed here, it should be noted that, apart from the unnecessarily severe bottlenecks which were created in these two sectors due to the high levels of government spending, such an allocation meant a movement of resources away from agriculture and industry and into urban-based roads, construction and symbolic projects; this resulted in the lower availability of resources for the sectors producing traded goods. The move into big projects, housing and transport was therefore at the expense of small projects and other sectors, in spite of the large increase in government expenditure. In fact, as will be discussed in Chapter 3, agriculture and especially the small-scale operators were starved of funds while what little was allocated to agriculture was spent on large and expensive projects.

Such a political allocation of resources results in a double squeeze of traded goods: first, through the process of real appreciation induced by the level of expenditure which results in relative price fall of the traded goods producing sectors, and second, through an inappropriate pattern of public expenditure which denies the traded goods sector access to resources and technology necessary for raising its levels of productivity. Therefore the analysis of adjustment must always take account of not only the aggregate level of government expenditure but also the components and the process through which the expenditure is brought about – and this is largely institutional. Hence institutional factors enter the analysis significantly since the planning, budgeting, and appraisal procedures do have an important impact in affecting the final supply response to changes in relative prices.

In any case, the quantum jump in expenditures proceeded. This meant a rapidly expanding role for the state in the economy. During the 1970s public consumption expenditures rose in real terms by 12.2 per cent per year, this was among the highest rates in the world.

Private consumption expenditures in real terms between 1973 and 1978 grew by almost 10 per cent per year comparable only to other oil exporters or countries with large foreign exchange inflows (World Bank 1983, p. 154), and 5.4 per cent per year between 1970 and 1980 (see Table A.6). Private consumption expenditures went into absolute decline after 1978.

Gross domestic fixed capital formation, rose at 11 per cent per year in the 1973–8 period, or 8 per cent per year if we take the 1970–80 period. Again, 11 per cent per year in real terms is a very rapid investment growth rate, especially given the rapid growth of consumption in the same period (Table A.6).

The rise in domestic expenditure meant a rise in demand for both imports and non-traded goods. Aggregate merchandise imports grew at an astronomic rate.

Table 2.3 Total merchandise imports (billions of dollars)

1970	71	73	75	77	79	80	82
2.0	5.0	10.6	15.9	16.5	11.5	15.74	13.4

Source: Bank Markazi Iran, *Annual Report*, 1972, 1976, 1979, 1983.

Table 2.3 shows the rise in imports between 1970 and 1982. By 1977 imports had risen to over eight times their 1970 level. The tremendous growth of imports between 1973 and 1977 is quite consistent with the adjustment mechanism described in Chapter 1, in terms of both the rise in demand for traded goods and the sheer pressure which must have been exerted on import-related domestic infrastructure and services (non-traded goods) with the consequent rise in their relative price (the extent of the relative price rise is discussed in the next section), though the stage one analysis is clearly insufficient since it ignores the critical role of the government. Tracing the figures until 1982 shows that the previously mentioned fall in oil exports during the 1978–80 period was not matched by an equivalent fall in imports, again indicating a government decision not to adjust to falling foreign exchange reserves and to finance the previously high levels of imports.

It is interesting to note that it is not only the level of imports but also the structure of merchandise imports which changes. Figure A.1 shows the changes in the different classes of imports. What is evident is the increase in the share of consumer goods in total imports, from 10.5 per cent in 1971 to 26.4

per cent in 1979, and the proportionate increase has been steady. The share of capital goods has a cyclical decline and the share of intermediate goods also has a falling trend. In other words, the import boom was associated with the increase in the share of consumer goods in the total import bill. The implication of such a shift for the industrialisation process of Iran is discussed more fully in Chapter 3.

Now consider the non-traded goods sector. Oil-financed expenditure is largely inflationary since the generated demand for non-traded goods has no equivalent domestic supply in the short run, hence raising the prices of non-imported goods. Throughout the boom period, serious bottlenecks were created as a result of the rapid rise in expenditure and the shortage of key non-traded services. Consider port capacity as an illustrative point. At the beginning of the Fifth Plan, port capacity was 3.8 million tonnes and the plan envisaged a capacity of 9.8 million tonnes by 1978. The revised version of the Plan increased requirements to 29 million tonnes by the end of the same year (PBO 1975, p. 297). The time-lag in delivering this capacity was obviously not taken into account and it was assumed that the allocation of a large budget for this purpose would suffice.

The import upsurge referred to above also created havoc in the southern ports. Graham (1970, p. 93) writes:

> At Khoramshahr, the principal port, over 200 ships were waiting to unload their cargoes by mid-1975; ships were having to wait 160 days and more before entering the harbour. At one point more than one million tons of goods were being kept in ships' holds awaiting the opportunity to unload... 12,000 tons were being unloaded per day but only 9,000 tons were being removed per day. At the most congested point in September/October 1975, there were over 1 million tons of goods piled upon the jetties and around the port.

Emergency remedial measures were adopted by the government. Large numbers of trucks were purchased which, instead of reducing the pressure at ports, aggravated the problems since it meant additional pressure on port capacity and on the demand for drivers. Shortage of skilled and unskilled manpower resulted in ill-conceived and hurried imports of labour which created additional difficulties, especially for housing. Put simply, the government always resorted to creating additional supply instead of attempting to reduce demand.

With rising rents, workers demanded higher wages which led to greater rural–urban migration into the poorly prepared cities. Also, rising land values increased the cost of public projects which in turn resulted in further injections of money by the government into the economy.

Consumer prices in the 1973–8 period on average rose officially by 12.8 per cent per year. For the 1976–7 period caution is needed regarding the price figures since official prices appear to have been used in constructing the price indexes. It is important to add that rising prices of imports do not account for the rate of inflation since the rate of inflation of imported prices at the same period was about 6 per cent per year (for example the US Consumer Price Index from Table A.7, the United States being a key trading partner) and also some of the important imported goods such as wheat were heavily subsidised so that the rise in their prices would hardly affect the price of such items inside the country.[9] Moreover, from mid-1975 to late 1976 there were officially announced price ceilings for a number of consumer goods, but the official price was much below the actual price which the same goods fetched in the black market (discussed below), and therefore the official consumer prices underestimate the actual rate of inflation for these two years (1975–6)[10]

Since more emergency high-cost projects did not seem to break the bottlenecks and lessen inflation, the government resorted to direct price fixing. In early August 1975, the Shah announced a price-control policy aimed at restoring the prices of a large number of commodities to their pre-oil-boom levels. An official price list was published and university students were hired to go out into the market place as inspectors. In the first two weeks some 10,000 persons were arrested, among whom were some of the most prominent industrialists (*Keyhan International*, 8 August 1975). The Minister of Commerce announced that by the end of the year Iran's inflation rate would be zero(!).

The only 'benefit' of the price-control programme turned out to be a decline in the official inflation indices since all government agencies were ordered to use the published lists. Apart from the official deflation, the price-control programme was a total failure, particularly as it resulted in a decrease in production and black-market activity. Shortages became more evident and investor confidence in Iran was substantially eroded.[11] Indeed the backbone of the economic success of the 1963–73 period which was the new entrepreneurial class became frustrated and alienated from the regime.

The boom, however, did not last beyond 1976. Its undoing was not caused by internal factors but from a drop in the income from oil due to the decline in oil sales accompanying the recession in the Western economies. Oil sales for the 1973–8 period had been anticipated to be 6,628.5 billion rials ($ 98.2 billion) while actual sales realized were about 20 per cent less than this (Razavi and Vakil 1984, p. 90). The overall budget deficit for 1976 alone was 8 per cent of expenditure. For the 1976–8 period the actual overall budget deficit was 13.5 per cent of actual expenditures.[12] Iran was now a borrower in international markets. It was no longer possible to meet the development targets and there was a belated recognition of the dangers of overspending. An Imperial Commission was set up to investigate waste, inefficiency and corruption. The Shah had decided to put the blame on an inefficient and corrupt bureaucracy. A bureaucracy which had been ordered earlier to indulge in a spending spree was now blamed for doing so.

The slowdown in government expenditure had to wait until 1977 because of the Shah's stubbornness. By then there was no choice but to launch a deflationary programme, though it was already too late to save the regime. A new prime minister took office and a new cabinet was formed with the primary responsibility of bringing the economy back under control and balancing the government's books. The turnaround came suddenly and abruptly with dire consequences for both the economy and the regime. A rapid rise in unemployment and severe shortages brought about an economic environment which laid the foundations for the overthrow of the Shah.

How does the above discussion on emergency government actions relate to the main argument of this book. The main argument of the book, as will be recalled, is that the 'automatic adjustment mechanism' is insufficient as an explanation of the economic response of the economy to the external shock of the oil price rise and does not explicitly deal with compositional effect of higher expenditure especially since government's macro and sectoral policies can substantially affect the economic outcome. In essence, the argument of the book is both a call for disaggregating the income effects of the oil boom and making explicit the impact of composition of public expenditure since the government has such a major role especially in the initial round of increases in expenditure as well as looking at the institutional environment in which higher expenditure comes about. It is the government's action which determines the percentage of extra income accrued to different sectors of the economy and one cannot take the Dutch-Disease model as if the additional income was directly accrued to the private sector, as in a coffee or gold boom. The relevance of the discussion on hurried planning and emergency measures is to show how government's misintervention in the economy affected the adjustment process. In Chapter 1, two modes of government intervention was identified — the 'planning mode' and the 'crisis mode'. The emergency measures are typical examples of the crisis-mode type of intervention by the government in the economy. The planning process became more planning in crisis and government's remedial measures to bring the economy under control added fuel to the fire.

The exact impact of such crisis-mode interventions is difficult to quantify. Nevertheless, one can arrive at indications of the largely negative impact of such measures by a number of examples. The hurried planning resulted in massive amounts of waste and misallocation of resources; the price-control programme or rather the 'price war' resulted in the erosion of confidence in the regime and the frustration of the entreprenurial class. For industrial capital in particular, these arbitrary actions on price controls were, as will be demonstrated in the next section, combined with falling domestic terms of trade.

Shortly after the introduction of price controls, severe shortages appeared on the scene. Daily consumer items now went onto the black market and were being sold at several times their official price, sharply raising the cost of living especially in the urban areas.

Having broadly analysed the policy environment, we can now proceed by looking at the non-oil GDP and its components. The actually observed output is of course the result of both economic and policy forces and the exact weight of each has proven difficult to quantify.

The non-oil Gross Domestic Product and its components

This section discusses the observed sectoral developments (output and prices) within the broad framework which was outlined in Chapter 1. An important question here is the extent to which the terms of trade of each sector can explain its output performance and the extent to which other factors (such as policy) may be involved in the determination of production. First, the non-oil GDP as a whole is discussed. Second, the changes in relative prices of the four main sectors of the economy – agriculture, manufacturing and mines, construction, and services – are analysed. Third, the price of labour in different sectors of the economy is discussed. Fourth, the growth of real output in the main sectors is discussed in order to see the extent to which real output changes follow relative price changes. Fifth and last, there is an analysis of the changes in the sectoral distribution of the GDP and some of its implications.

The non-oil GDP

Continuous trend growth rates are estimated for the non-oil GDP and its components for the three periods 1962–72, 1973–8, and 1973–80, in order to separate the period of the oil price rise from the 1960s, as well as from the post-1978 events (Table A.10). The annual trend growth rate of real, non-oil GDP rose from 9.1 per cent in 1962–72 to 9.8 per cent in 1973–8, subsequently falling so that over the 1973–80 period, the growth rate was 6.9 per cent per year. It should be noted that the growth of the post-1973 period is more unstable than the growth of the earlier period. The R^2 associated with these trend growth rates is lower in the 1973–78 and 1973–80 period, indicating greater fluctuations around the mean. The higher and the more unstable growth of the real non-oil GDP after 1973 was also accompanied by a much higher rate of inflation. The annual inflation as measured by the non-oil GDP deflator rose from 2.1 per cent in the 1962–72 period to 14 per cent in the 1973–8 period and 13.7 per cent if we take the 1973–80 period to 14 per cent in the 1973–8 period and 13.7 per cent if we take the 1973–80 period.

Changes in relative prices

Comparing the 1962–72 period and the post-1973 period first, it should be noted that the former period is one of general price stability while the latter is one of general price inflation. Using the sectoral GDP deflators for comparing relative price changes, it can be seen that, in the earlier period, construction has the most favourable relative price change while the change in the GDP deflator for the agricultural sector is in line with that of inflation in the non-oil GDP. Manufacturing, mines and services, however, experience a slight decline in their price relative to the other two sectors (Table A.10).

In 1973–8, again construction has the most favourable change in its terms of trade, the GDP deflators for agricultural commodities and of services remain close to the average inflation in the economy, while prices in manufacturing and mines continue to deteriorate The change in the terms of trade may be better illustrated if we look at the cumulative rise in prices between 1973 and 1978 – these are given in Table 2.4 , along with figures for the 1962–72 period. Table 2.4 shows the cumulative changes in the GDP deflators for the main sectors of the non-oil GDP.

Table 2.4 Cumulative percentage price increase of main sectors

	1962–72	1973–78	% of GDP 1973–78
Agriculture	37.7	88.0	40
Manufacturing and mines	22.4	69.7	12
Construction	70.0	216.7	4
Services	10.5	98.4	35

Source: Calculated from Table A.9.

Caution is needed in drawing strong inferences regarding the exact rate of real appreciation given the underestimation of price rises by the official data during the 1976–7 period, although the main trend and price relatives are unlikely to have been substantially different. Table 2.4 shows that construction prices had the fastest relative price growth in both periods while the relative prices of manufactured goods declined relative to construction and agriculture but rose relative to services in the earlier period; in the later period, however, the relative price of manufactured goods fell with respect to all other sectors. Agricultural prices, which in the earlier period were rising faster than both manufacture and services, drop to the third place in the later period being ahead only of the prices of manufactured goods. Finally, prices of services rose least in the earlier period and accelerated considerably in the later period. So the contrast between the two periods is sharp, especially with regard to services. The acceleration of price increases of construction and the sharp rise in the price of services are consistent with the economic analysis presented in Chapter 1, since these are both sectors with a non-traded product although the extent of traded inputs in these two sectors is likely to have reduced their relative price rise while the extent of non-traded inputs is likely to have increased their relative price rise. Over 88 per cent of construction activity in Iran in the 1970–79 period has been for residential purposes (*Statistical Yearbook of Iran* 1980–1, p. 754). In terms of inputs, the main material for residential houses is iron poles and bricks, the latter being a non-traded good. Labour costs are the other major cost item on the input side (the correlation between wages and growth of output in the construction sector is analysed on pp. 50–1 below). Labour remained mostly non-traded, although there was some import of unskilled labour from neighbouring countries. Land is, by definition, non-traded. In other words, construction is likely to have experienced rapidly rising costs.

The rapid rise in expenditure and of the price level result in the appreciation of the real exchange rate in Iran. The real exchange rate can be measured by correcting the nominal exchange rate by the differential rates of inflation between countries which are trading partners. Table A.7 shows the changes in the exchange rate between Iran and the United States for the 1968–80 period. It shows that between 1973 and 1978 the rial appreciated by almost 44 per cent against the US dollar while the nominal exchange rate, as was shown earlier, depreciated slightly against the dollar in the same period. For the 1973–8 period the real exchange rate appreciated by 35.3 per cent.

Before proceeding to analyse the behaviour of real output in the light of the above relative price changes, it is necessary also to discuss the changes in the price of labour or the wage rate.

Price of labour

The price of labour, although relatively stable in the 1960s, rose rapidly in the post-boom period. Wages and salaries in construction rose at 27.1 per cent per year, in the manufacturing sector at 27.6 per cent per year, in agriculture (using the wage in the construction sector as the opportunity cost of rural labour) at 27.1 per cent per year; the wages and salaries of the government sector rose only 13.3 per cent per year but renumeration in the government sector does not reflect market forces and in any case there are fringe benefits involved in being employed by the government which are not reflected in the index presented here (Table A.11). Therefore, using the homogeneous labour assumption of Chapter 1, it will be assumed that nominal wages grew at about 27 per cent per year for the economy as a whole which implied an average annual growth of 13 per cent in the real wage if the non-oil GDP deflator is used or a real wage growth of 10 per cent if the CPI is used for deflating the nominal wage rates. GDP deflators for the main sectors of the GDP for the 1973-78 period had the following annual trend growth rates: construction, 29 per cent, manufacturing, 9.7 per cent, services, 13.3 per cent and agriculture, also 13.3 per cent. Comparing the growth of prices and wages the following picture emerges:

$$P_c\!\sim\, >\, W\!\sim\, >\, P_s\!\sim\, =\, P_a\!\sim\, >\, P_m\!\sim$$

where $P_i\!\sim$ denotes the percentage change in the GDP deflator for the ith sector (construction, services, agriculture and manufacturing), and $W\!\sim$ denotes the percentage change of wages. In other words, the real wage rose relative to agriculture, manufacturing, and services but it fell relative to construction. The cost of housing must have become an increasingly major component of the household expenditure, assuming a relatively price-inelastic demand for housing and the responsiveness of rents to construction costs, including the cost of land, which rose rapidly in this period. The rise in W/P_t means that traded goods (especially food) are likely to have been a major component in the total consumption basket of the workers in Iran. Moreover, the rise in the real wage relative to agriculture reflects the cheap-food policy which was followed in Iran in this period (more on this in Chapter 4).

For construction, agriculture, and manufacture the picture is basically in conformity with the expectations of the model in Chapter 1 where, due to the implied magnification effects, rents in the non-traded sector would increase by more than wages and wages would increase by more than rents in the traded sector, that is

$$P_c\!\sim\, >\, W\!\sim\, >\, P_a\!\sim\, >\, P_m\!\sim.$$

The only sector where wages and prices do not conform to the model is in the service sector, since services are usually assumed to be non-traded, but wages (in non-government services) rose faster than the GDP deflators for services. Part of the explanation for the relatively slow price rise of services is the large share (about 35 per cent) of total value added in services being due to government services. The GDP deflator for services rose only as much as wages and salaries in the government sector or 13.3 per cent per year.[13] Also services must have benefited from the presence of traded inputs, for example, in transport. The transport sector was also highly subsidised and this prevented price rises in that sector. The presence of subsidised services and the large share of government services in the total value added in the service sector modifies the prediction of the inequality presented in Chapter 1, since $P_s\!\sim\, <W\!\sim$. This is also a point which further demonstrates the impact of government policy upon the relative price predictions of the Dutch-Disease model.

Sectoral growth of non-oil GDP

There is a sharp contrast in the sectoral growth of the non-oil GDP between the 1962–72 and 1973–8 periods. In the 1962–72 period, manufacturing and mines is the fastest-growing sector, at 12.3 per cent per year in real terms, followed by services which grew at 11 per cent, construction at 7.6 per cent and agriculture at 4.2 per cent. In the following 1973–78 period, on the other hand, manufacturing and mines, and agriculture are the lagging sectors while services and construction on the whole achieve rates of growth which are above the average of 1962–72 period. The contrast is quite clear, especially looking at manufacturing and mines, which experienced a sharp fall in average growth of real output. In other words, manufacturing lost its position as the fastest-growing sector in the 1970s (Table A.10). Agriculture's growth (trend) also slowed down to 70 per cent of its original growth rate with sharper fluctuations around the mean ($R^2 = 0.67$). Principal factors behind the slowdown of the growth of the agricultural sector are discussed in more detail in Chapter 4.

Concerning the manufacturing and mines sector, it is important to distinguish between the 1973-76 period and 1977 onwards. In the first period, the growth of manufacture and mines was of the order of 16.7 per cent per year while, from 1977 onwards, the sector showed substantial negative growth rates reflecting factors such as the very high growth in wages relative to productivity, power shortages, political uncertainty, and worker agitation in the factories during 1978 and 1979. Considering the 1973–8 period as a whole, however, the manufacturing sector is a lagging sector relative to construction and services, although some sub-sectors of the manufacturing sector fared considerably better than others, a point which will be discussed in the next chapter.

In the same period, services had a continuous and relatively stable growth of 16 per cent per year and even if we include 1979 and 1980, the growth of services still remains at a level of over 12 per cent.

Construction grew at 10.9 per cent per year, which was faster than its growth during the 1960s, but its growth was rather unstable, with $R^2 = 0.49$ for the 1973–78 period. Between 1973 and 1978 the construction sector had three years of positive growth and three years of negative growth. In 1979 and 1980 also, construction had negative growth and there was a decline in gross fixed capital formation in the construction sector after 1976 (Table A.12). Moreover, if the percentage change in the price index of construction is compared with the percentage change in wages on an annual basis, a very interesting correlation emerges. Table 2.5 shows a remarkable positive correlation between the growth of output in the construction sector and the relative rise of prices over wages in that sector ($r = 0.89$). In all the years when prices rose faster than wages, output grew. In all the years when wages rose faster or even equal to prices, output or net value added declined. Although this relationship is not intended to show any rigid relation between these two variables, it broadly validates the theoretical model of Chapter 1 and the expected growth in the output of the non-traded sector due to prices rising faster than wages in the sector. The role of wages becomes additionally important when it is realised that, between 1970 and 1979, on average over 88 per cent of all domestic fixed capital formation in construction was in residential buildings, which are mostly private and highly labour-intensive (Table A.12). The sharp rises in construction output in 1973, 1975, and 1976 was an important source of rising demand for labour and thereby of rising wage-rates.

Table 2.5 Wages and prices in construction

Year	% change wages	% change in prices	sign of change in real output
1972	17	20	+
1973	21	24	+
1974	28	28	−
1975	47	68	+
1976	39	56	+
1977	34	20	−
1978	19	20	−
1979	15	11	−
1980	19	20	−

Source: Tables A.9 and A.11

Other input costs such as the price of land, as well as policy variables such as construction permits, also play an important role in determining output growth. Unfortunately, it has not been possible to obtain an index of the rise in land prices, although by unofficial accounts the price of land rose astronomically in the 1970s and this must have been a further factor in determining the rate of construction growth since higher urban land prices would increase the cost of construction and thereby lower profits in that sector. In other words, with regard to construction costs, two non-traded factors enter into the explanation: these are the labour-cost effect and land as a specific non-traded factor effect, both of which would lower profits in the construction sector. At the initial round of price increases, the price of construction goes up and pulls profits up in that sector. Higher derived demand for construction inputs in turn leads to a rise in wages and land prices which consequently lowers profits in the construction sector.

Changes in the sectoral distribution of GDP

The sectoral pattern of the non-oil GDP also changes significantly. It is GDP and not GNP which is the focus here because as has already been said, the book is an analysis of the structure of the domestic output and not of the income received by domestic residents. It should be added that one cannot attribute all the changes in the sectoral growth of the GDP to the impact of higher oil revenues since presumably structural change would have taken place in the absence of higher oil prices. Nevertheless, it is the impact of oil revenues on the direction and speed of structural change in an oil economy which is the concern here.

Looking at the shares of the various sectors in the GDP in the same period (Table A.13), the two outstanding features are the decline of the share of the agricultural sector from over 20 to under 10 per cent of the GDP and a rapid rise in the share of services from 23 per cent to almost 50 per cent of the total GDP. The share of manufacturing and mines grows slightly from 7.7 per cent in 1962 to 11.8 per cent in 1976 and 9.5 per cent in 1980, while the share of construction is relatively stable at around 5 per cent of the GDP. The inverse correlation between the decline of the share of agriculture and the rise in the share of services in the GDP reflects the rapid reallocation of labour away from agriculture and towards services in the past two decades in Iran. The structure of non-oil GDP also reveals a similar but sharper trend (see Table 2.6).

Table 2.6 Percentage distribution of non-oil GDP (1974 prices)

	Agriculture	Manufacturing	Construction	Services
1962	38.0	13.0	8.2	40.8
1968	29.6	16.0	8.9	45.5
1973	20.1	18.5	7.1	54.3
1974	18.6	19.1	6.0	56.2
1977	13.7	15.1	7.5	56.1
1980	15.1	10.2	4.9	69.8

Source: Tables A.8 and A.9

Changes in the structure of the non-oil GDP indicate that it has been the service sector which has been the main growth point of the economy. All the other sectors have shown a declining trend of structural shares.

Conclusions regarding non-oil GDP

So far, the evidence derived from the analysis of the GDP suggests a broad confirmation of the economic model presented in Chapter 1, although there are a number of points concerning supply behaviour which require further investigation. As far as concerns construction and services becoming the fastest-growing sectors and agriculture and manufacturing being the lagging sectors, the evidence agrees with the hypothesis although the service sector does not confirm fully in terms of price rise. The inequality $P_n \sim > W \sim > P_t \sim$ appears to have become

$$P_c \sim > W \sim > P_s \sim = P_a > P_m \sim,$$

that is, construction prices rose faster than wages and prices of the service sector and agricultural sector, which in turn rose faster than the prices of the manufacturing sector. The rise in wages relative to the price of traded goods (agriculture and manufacturing) is in accordance with the hypothesis. Wages fell in terms of construction but not in terms of all non-traded goods since they rose faster than the price of services.

The service sector does not conform fully in terms of price rise if it is assumed that services are non-traded, since prices in this sector, although rising faster than in manufacturing, did not grow faster than in agriculture. This is partly due to the important share of government services and government subsidies on transport. Data on factor intensity of services in Iran is rather scanty and it is difficult to know to what extent rising labour cost would have affected that sector.

To investigate further the factors underlying the supply response, in addition to analysing the relative roles of both economic and policy variables particularly in relation to the traded goods sector, we now proceed in the next two chapters, to analyse the manufacturing and agricultural sectors in more detail.

Notes

1. Calculated from Table A.8. The share of oil in GDP at *current* market prices is the relevant figure for giving the appropriate order of magnitude since it was a *price* change and not a quantity change which was the main force behind raising the share of oil in the GDP. The level of real activity in the oil sector is assumed to be exogenous in the present analysis since oil has little direct participation in the economy apart from its impact through the fiscal linkage. In fact, taking the real share of oil and gas in the GDP, the rise in the share of oil between 1969 and 1974 is only 3 per cent, reflecting the fact that it was not the increase in oil production but oil price which increased its share in the nominal GDP. The 1982 figure is taken from the latest Annual Report which was available too late to be included in Table A.2.
2. Calculated from IMF (1983, p. 281) The rial's movements against the SDR reveal a similar trend.
3. The reason for the difference between our results and the results of Jakubiak and Dajani (1976) lies in the difference in the figures concerning bank credit to the private sector. Jakinbiak and Dajani use figures for bank credit which are consistently much less than the figures in the BMI's *Annual Report* which we have used. However, our estimates of the balance-of-payment deficit of the private sector is the same as the above study. It is difficult to believe that the IMF has a more correct estimate of bank credit to the private sector than the Central Bank of Iran (BMI). It is therefore reasonable to believe that the 1976 study has incorrect figures concerning the amount of credit to the private sector and, hence, arrives at the incorrect conclusion that the private sector has had a contractionary monetary effect.
4. Using a formulation with lags where the money supply in period u is a function of the other two variables in period $u-1$ produces insignificant t-ratios and a lower R^2.
5. There are two studies of the demand for money for Iran with similar conclusions, namely that there is a stable income elasticity of the demand for money in Iran and its value is in the range

the range 1.4 – 1.6. These studies are Crockett and Evans (1980); and Morgan (1979). We compared the actual money stock and the estimated money demand figures to calculate percentage change in 'excess cash balances' and compared it with the percentage change in the consumer price index both in the same year and using a one-year lag. The correlation between these two variables is of the order of 0.1, that is, there is hardly any correlation between the growth of excess balances and the rate of inflation. The evidence, however, is not inconsistent with the idea that money supply, especially since 1973, grows faster than money demand, even though the demand for money may not have been a stable function of income. In fact, it may be argued that the period since 1973 has been essentially one of monetary disequilibrium with supply-determined as opposed to demand-determined money balances.

6. The following account of the experience of the Plan and Budget Organisation is based on the same source. Absorptive capacity here especially refers to port capacity, transport and manpower.
7. Military expenditures were around 30 per cent of the total budget and 50–60 per cent of the current expenditures in the 1970s. In some years military expenditures would even exceed the legal limit of 30 per cent by reclassifying certain military projects as development expenditure. The actual military expenditures are available from the Statistical Yearbook of Iran, 1980–81, p.733.
8. One example is the huge imitation of the Eiffel Tower in the middle of the square which leads to Tehran Airport.
9. Relative price movements are analysed in detail in the next section.
10. Unfortunately open-market prices for these two periods are not available and this book has relied on the official GDP deflator for analysing relative price movement for the entire period including these two years. Therefore, caution is needed in drawing strong inferences from the data for these two years.
11. This point will be referred to again in Chapter 3.
12. Calculated from Table A.3.
13. The share of government services in total value added in services is from Bank Markazi Iran, *Annual Report and Balance Sheet*, 1976–77, p. 12.

3 The manufacturing sector in Iran

The experience of the manufacturing sector in Iran will be analysed by considering the impact of the post-1973 rise in expenditure on the industrial structure. First, we discuss some of the general features of the rapid industrialisation of Iran during the 1960s and early 1970s. Second, we discuss price differences between the domestic ex-factory prices and the CIF import price, and will both demonstrate the relative efficiency of these sectors and indicate the impact of the government's tariff policy, import controls, and direct price controls. Third, we draw attention to an important difference among manufacturing activities, namely, the extent to which they rely on domestic resources for their imports, as opposed to imported inputs, hence determining the share of traded goods in their input structure.

Fourth, we introduce and discuss the impact of higher real exchange rate both on the manufacturing sector as a whole and on different types of manufacturing activity. We show that the overall impact has been detrimental to manufacturing activity, although the impact varies among branches of manufacturing. Among the key factors relevant in explaining this differential impact are the ratio of non-traded inputs to total value added, the degree of price protection granted to the industry by the government, and access to official credit by the firms. In terms of inputs, industries whose inputs are largely based in domestic resource will, *ceteris paribus*, be more pressurised than industries which rely on imported inputs since domestic resources contain a greater element of non-traded goods and services than imported inputs. In other words, the greater the ratio of non-traded inputs to total value added or the greater the share of non-traded inputs in the total cost of production, the higher the adverse impact of real appreciation on the industry's value added. For the empirical analysis in this chapter, domestic inputs are used as a proxy for non-traded goods since it has not been possible to find data for specific non-traded inputs on an aggregate basis. On the output side, those industries most exposed to international trade would experience the smallest price rises. The extent of exposure to international trade is affected by both transport costs and government's commercial controls and pricing policy.

It should be added that, since the post-1973 period was a boom situation, most industries expanded, and that it is difficult to provide conclusive proof of the differential impact of the boom on industries. The analysis in this chapter is, therefore, more suggestive than conclusive. An important point which is emphasised is how the resource-based industries which had a comparative advantage prior to the boom were, in fact, squeezed more than import-based industries and that this has damaged the foreign exchange earning capacity of the non-oil economy. This analysis is an attempt to suggest that the economic policy of high levels of domestic expenditure and increases in the prices of non-traded goods can create short-term price signals based on temporary changes in resource endowments which distort relative prices against those industries with long-term comparative advantage, that is, those which are least dependent on imports for their inputs. Furthermore, it is shown how government policies towards the industrial sector further worked to the detriment of industries with traded products prior to the oil boom.

Industrialisation in Iran, 1962-72

To begin the analysis, it is necessary to consider industrial development prior to the oil boom and how the adjustment process affected the industrial structure inherited from the 1960s.

Prior to 1973, Iran had already achieved a considerable industrial base with manufacturing and mines accounting for 9.1 per cent of the GDP and 18.1 per cent of the non-oil GDP, an industrial base which was largely built up during the 1960s. Following a number of reforms in the early 1960s which were to a large extent in response to a balance of payments difficulty, Iran adopted an import-substitution strategy of industrialisation.

There are a number of distinctive features of industrial growth in this period. First, industrialisation proceeded at a very rapid and consistent pace. Gross value added in manufacturing increased at an average annual rate of 12.3 per cent during 1962–72 in real terms (Table A.10).[1] The growth in real value added also accelerated towards the end of 1960s, with an annual average of 12.6 per cent during the 1968–72 period. Moreover, the share of manufacturing activity in the GDP increased from 7.7 per cent in 1962 to 8.7 per cent in 1972. The share of manufacturing in the non-oil GDP rose from 13.4 per cent in 1962 to 18.2 per cent in 1972.[2] The contribution of manufacturing to the incremental non-oil GDP was over 21 per cent.

Second, in the 1962–72 period there was parallel growth in light consumer goods, intermediate goods, and consumer durable and capital goods. The growth in light consumer durables (food processing, textiles, and footwear) was 10.7 per cent, intermediate goods (steel, fertilisers and chemicals) 19.3 per cent, and consumer durables (radios, televisions and private motor cars) and capital goods (cement and electrical equipment) 21 per cent. Their shares in incremental value added were 33.2, 31.3, and 41 per cent, respectively (IBRD 1972, p. 56). Most of the growth in non-durable consumer goods occurred in processed food, clothing and textiles. In intermediate products, basic metal industries had the fastest growth, rising in real value added from 10 million rials in 1962 to 6.5 billion rials in 1972. In consumer durable and capital goods, most of the growth was attributable to the motor-vehicle industry and electrical equipment.

Third, manufacturing growth was largely based on the home market. Industries simply grew in response to an existing home demand which was until then satisfied by imports. The share of manufactured exports in total manufactured value added was 13.5 per cent in 1960 and about the same in 1970. Although small, the value of manufactured exports rose at an average rate of 10.5 per cent from 1962 to 1970 (Table A.16). Moreover, in this period the share of textiles exports (mostly carpets) declined, while other non-traditional manufactured exports rose in both share and volume (Table A.16).

Fourth, while most of the industrial growth took place in the private sector of the economy, the government exercised considerable influence over industrialisation mostly through tariff policy, import controls, fiscal and financial incentives, and licensing. Let us consider these factors in more detail.[3]

Tariff protection was highest on consumer goods and consumer durables, less on intermediate goods and least on capital goods. Average *ad valorem* tariffs for various products are shown in Table A.17. The range of tariffs for consumer goods varied from 65 per cent for food products to 45 per cent for beverages. Tariffs on intermediate goods ranged from 56 per cent for paper and printing to 137 per cent for non-metallic products. In the capital goods group, the rates were 59 per cent for machinery and 104 per cent for transport equipment. The average duty levels for the consumer, intermediate and capital goods are respectively, 135, 95, and 79 per cent. Moreover, tariffs generally rose in the 1960s and by 1970 the equivalent duty was on average four times the 1960 level (Table A.18). Furthermore, quantitative restriction on imports were heaviest on consumer durables, less on capital goods, and least on intermediate goods.

In order to understand the divergence between domestic and import prices, however, it is not sufficient to look at nominal tariffs. The appropriate measure must also take into account the impact of import restrictions, direct price controls, and tariff redundancy caused by domestic competition. One way of arriving at the actual protection received by domestic producers is to use the domestic ex-factory price and compare it to the CIF import price which would indicate the combined impact of price measures, non-price measures and domestic competition. This method, of course, is not without shortcomings. Comparisons must be made between similar products, which is difficult to verify without close inspection. Also the CIF import price may be subject to transfer pricing, and consequently not indicating the long-run import price. Moreover, the exchange rate used to convert the domestic price to a border price may itself not reflect the true scarcity value of foreign exchange. With these limitations in mind, let us proceed by looking at such price comparisons.

The World Bank carried out a study of price relations in Iran in 1970. Data were collected on 43 products through plant surveys and CIF unit prices of similar goods obtained from foreign trade statistics. In a number of cases it was found that the import price net of freight was significantly lower than the price in the country of origin, which indicates the presence of transfer pricing and/or subsidies. Table A.19 shows the implicit duty (protection received) and the nominal duty level for a number of products.

In most cases there is a considerable difference between nominal and implicit tariff. The excess of domestic price over the CIF price was in most instances lower than what is implied by the nominal tariff. This phenomenon can be interpreted in a number of ways. Sadigh (1975) and Alizadeh (1985) have both interpreted this as tariff redundancy which largely reflects domestic competition and increasing efficiency of domestic production. Another possible interpretation is that the difference between the domestic and the CIF price reflects quality differences between the domestic and imported goods – the imported ones sell at a premium. This is an inplausible interpretation, however, as it assumes that there are no barriers on imports and the only factor causing the price differential is the quality difference. The presence of quality differences, if anything should increase the price differential between the domestic and the border price, not narrow it. Therefore, the first interpretation concerning the increasing competition in, and the efficiency of, Iranian industry in the 1962–72 period appears more plausible.

Price controls also played some role in preventing the domestic products from reaching the price implied by the tariff. Table A.19 also shows that some home products were even competitive with imported goods at CIF prices. On the whole, out of 39 products in Table A.19, 25 have an implicit tariff less than the nominal tariff. For nine products the implicit tariff was greater, reflecting the importance of quotas, and for five products the nominal and the implicit tariff were almost equal. Moreover, looking at the price trends in the 1962–72 period, the data suggest that the prices of manufactured goods in the early 1960s started from a high level, since importer margins prior to import-substitution had been high. From 1962 to 1972, however, the prices of manufactured goods rose less than most other categories of GDP (see Tables A.9 and A.10). Again this may be interpreted either as increasing efficiency at given factor proportions, resulting from economies of scale, or from increasing capital intensity and an outward shift in supply. It is impossible at this point to determine the exact share of each of these factors in affecting prices and quantities in the industrial sector in Iran during this period. What is of relevance for the argument of this book is that rapid industrialisation occurred in spite of falling relative prices both in terms of domestic prices and in terms of border prices, and this reflects considerable progress towards establishing a viable industrial base. This is an important argument for establishing the central argument of this chapter – that the following post-1973 period was essentially a reversal of this process and reflected a backward move, as far as establishing a viable industrial base is concerned.

Now consider the extent to which domestic industry utilised domestic inputs.[4] Table A.20 shows the percentage share of imported inputs in the value of output. Local content can rise either if industry is labour-intensive (for example, telephone equipment) or when the inputs are of domestic origins (for

Table 3.1 Iranian industry: national economic profitability rating

A. High value added and relatively low prices

Cotton textiles	Cement
Footwear	Construction materials
Canned fruits and vegetables	Electric meters
Meat packing	Paper (?)
	Telephone receivers

B. Low value added and relatively low prices

Woollen textiles (?)	Tyres
Vegetable oils	Pumps
Pharmaceuticals	Diesel engines (?)
Cables	Electric switchgear
Paints	Buses

C. High value added and relatively high prices

Sugar	Steel wire (?)
Caustic soda	Radios (?)
Glass	Space heaters (?)
Telephone exchanges	Electric fans

D. Low value added and relatively high prices

Synthetic fibre	Trucks (?)
Refrigerators	Passenger cars
Air coolers	Radios
TV sets	Space heaters (?)

Source: Avramovic (1970, p. 24)

example, cement) or both (for example, textiles). The extent of utilisation of domestic resources is important when we consider the impact of the increase in the price of non-traded goods on manufacturing costs. Costs, along with the ability of the manufacturers to raise prices, will determine the impact of adjustment on this sector. Putting together the results of Tables A.19 and A.20 on prices and import dependence, we arrive at Table 3.1, which shows four categories of manufacturing activity in terms of value added and output price.[5] These four are industries with high value added and low prices, high value added and high prices, low value added and low prices, and low value added and high prices. From the national economic profitability point of view, those industries are most profitable which yield highest value at lowest resource cost: this means industries in which the share of imported inputs is lowest and whose prices are least above the world market prices.

From the national economic point of view, those industries with high domestic value added and low prices relative to the international price were the ones with the most comparative advantage. In other words, industries at the top of Table 3.1, such as cotton textiles, footwear, canned fruits and cement, were the best from the point of view of economic growth per unit of scarce resource used; they are the ones which are likely to be more viable since they are based on domestic resources.[6]

It will be shown in the next section that industries with the greatest comparative advantage in the earlier periods declined the most, while industries with the least comparative advantage and most dependent on imported inputs and protection grew the fastest. In this sense, there was not only 'de-industrialisation' in the post-1973 period but also perverse 're-industrialisation' due to the changes in the industrial structure towards protected and import-dependent industries. This is elaborated upon subsequently.

Having reviewed some of the important characteristics of industrialisation in Iran in the 1962-72 period, we can now analyse the impact of the boom in the 1970s on the received industrial sector.

The post-1973 boom and manufacturing

Reference has already been made to the rapid increase in government expenditure after the 1973 oil-price rise which through the multiplier effect induced a rapid rise in aggregate expenditure. The expansion of domestic demand far beyond the country's domestic supply constraint resulted in an acceleration of the domestic rate of inflation which generally squeezed the manufacturing sector, although some sectors fared better than others. By 1977, the strains on the sector were becoming apparent and growth in value added, which had shown a continuous rise since 1962, became negative and remained so at least until 1980 (Table A.9). Fixed capital formation in manufacturing which had shown continuous growth since the early 1960s, and between 1970 and 1973 had grown at 12.5 per cent per year in real terms, only rose 1 per cent in 1977 and declined continuously until 1980 (Table A.12).

The reasons behind the sharp decline of the manufacturing sector included spiralling wage costs, serious shortages in electricity, and the limited ability of manufacturers to raise prices due to both imports and government price controls. By 1978, civil disturbances culminating in the revolution further contributed to the decline in industrial output. From early 1974, imports became more liberalised and import restrictions on a large number of consumer goods and industrial raw materials were lifted or tariff rates reduced. In certain instances the government used the lifting of import controls as a threat to domestic producers. In May 1974 producers of home appliances were warned that they must cut prices by 10 per cent or else import controls would be lifted (Keyhan, 25 May 1974, cited in Sadigh 1975, p. 329) Also the rapid increase in government oil earnings presumably lowered the importance of tariffs as a source of revenue for the government.

Lifting of restrictions on foreign exchange transactions in December 1974 plus the law which required that manufacturing companies had to sell 49 per cent of their shares to their employees and the public, further discouraged industrial investment. Also, the manufacturing sector witnessed a large number of strikes during 1978 and 1979 which contributed to the further decline of manufacturing output in this period.

Between 1973 and 1976, manufacturing wages rose by 2.6 times (Table A.11). Between 1976 and 1978, they continued to rise by an average of 27 per cent per year. In 1979, they rose by 60 per cent and in 1980, by a further 18 per cent. Using the GDP deflator for manufacturing and mines, it becomes clear that the prices in that sector between 1973 and 1978, only rose by 9.7 per cent per year which was almost six percentage points below the GDP deflator for the economy as a whole. In other words, wages rose several times faster than manufacturing prices. The rapid rise in wages and the cost of production in general meant that the high initial prices which prevailed in the 1960s were gradually eroded. Using the index of output per worker as a proxy for productivity of labour, it is also clear that wages rose far faster than productivity. Between 1973 and 1979, nominal output per worker in large industrial establishments rose on average by 12 per cent per year (*Statistical Yearbook of Iran* 1980-81, p. 469); given the price rise of 9.7 per cent for the manufacturing sector (Table A.10), the real productivity gains were 2.3 per cent per year, which is clearly too small and could not have compensated for the rise in wages.

Now consider electricity, a typical non-traded industry, the shortages of which were a significant factor behind the decline in industrial output after 1976.[7] The reasons behind the shortages were related to both supply as well as demand factors. On the supply side, since electricity is usually a non-traded good, its supply requires the expansion of domestic capacity. Given the time-lag between the decision to expand capacity, installation of capacity, and

63

actual power generation, short-term shortages are inevitable unless adequately planned for and an excess capacity is built-up prior to the rise in demand.

On the demand side, given the highly income-elastic character of electricity, higher income resulted in higher use of electricity. Moreover, due to the government's price-fixing policy, the relative price of electricity declined over time, which contributed to further consumption of electricity. In fact, growth in the domestic household use of electricity accelerated after 1973, while industrial use fell sharply. Between 1973 and 1976, for example, the industrial use of electricity went down from 41.3 to 29.8 per cent of total electricity consumption while the domestic use rose from 16.8 to 28.8 per cent.[8] In other words, the industrial users of electricity were literally crowded out by domestic households to the detriment of manufacturing output.

This is a key point relating to the hypothesis of this book and this chapter since here it can be seen how a particular government's pricing policy can affect the supply response of the industrial sector, as an independent factor quite apart from the macroeconomic adjustment policy concerning the aggregate level of expenditure. Again it is shown here how one needs both macroeconomic analysis and more detailed sectoral-level analysis for understanding the actual supply performance. The fall in the proportion of total electricity consumption accounted for by the industrial sector could also be interpreted as reflecting lower industrial growth in this period and it is likely that this was also a contributory factor. Nevertheless, the repeated complaints of the industrial establishment, as documented in the Industrial and Mining Development Bank of Iran (IMDBI) reports (1975–6, 1976–7; and 1977–8) indicates an undeniable shortage of electricity facing the industrial establishments.

The next two factors, namely, the share-divestiture scheme and the lifting of restrictions on foreign exchange transactions jointly contributed to lowering manufacturing investment and output. The share-divestiture scheme was introduced in 1975 to provide both a sense of participation and financial gain to employees, although it did not relate to the control of industry. Nevertheless, the scheme was not well received by industrialists and the fears of the business community led to a sharp fall of investment at the end of September 1975 (Graham 1979, p. 95). The fears of closer control and regulation by the government and higher taxes meant that the industrialists looked for an outlet different from manufacturing investment, At the same time, the relaxation of exchange controls was announced which facilitated the transfer of profits and liquid assets abroad. The total capital outflow accounted for by the private sector between 1974 and 1977 amounted to nearly $2.4 billion (Bank Marzaki Iran, *Annual Report* 1977, pp. 146–7. The share participation scheme was also an important factor behind the withdrawal of a number of foreign companies.

It should be added that the decline in business confidence was not solely attributable to the share-divestiture scheme. The price controls conducted through the anti-profiteering campaign also played a big role. Entrepreneurs and merchants were identified as the cause behind higher prices, and in the first two weeks of the operation 7,750 persons were arrested. Among those arrested were two of the top industrialists. In fact, one of the industrialists had one of the most efficient and sophisticated industrial plants, which closed down after his arrest (BMI, *Annual Report*, 1977, p. 94). As a result of the campaign, although official prices went down for a few months, black market prices rose substantially. Thanks to the 'war' on prices, shortages became more acute.[9]

In sum, rapid increases in costs, acute shortages, especially of non-traded goods, and disincentive price and non-price measures by the government all contributed to depressing manufacturing output as a whole, particularly after 1976. The discussion here is intended to show the impact of government intervention on the output of the industrial sector, which is a traded sector. As was mentioned earlier, the crisis-mode intervention had an adverse impact on investor confidence in this sector. Also government policies of high expenditure created severe shortages for the industrial sector.

The industrial sector as a whole, of course, is a highly aggregate concept which hides the diversity among industries. Certain industries, indeed, declined more than others, partly due to competition from imports but, more importantly, due also to rising wage and material costs relative to their product prices.

A rough indication of differential growth and decline among industries may be found by the number of existing industries in a particular activity. This could also reflect concentration resulting in the survival of the fittest or those firms best able to adopt cost-reducing innovations usually under cost pressure. Chart A.3 shows the number of large industrial establishments between 1971 and 1980. Let us consider some of these industries briefly.

Textiles

As chart A.3 shows, there were tremendous variations in the number of industries in each classification. One striking feature of the figure is the sudden rise and decline of the number of textile and leather firms in 1974–5 and their continuous decline until 1979. Whatever the reasons behind the fluctuation of the number of industries between 1972 and 1974, what is of interest for the economic model underlying the argument of this book is the continuous decline of textile establishments after 1973. Looking at the total quantity of output of various sub-sectors of the textile group, it can be seen that, between 1972 and 1977, the production of cotton synthetic fibres and of woollen textiles both declined (Table A.21).

In both cases, production reached a peak in 1975 and subsequently declined. In the case of cotton/synthetic textiles, production in 1977 was even less than production in 1972, while imports had risen from 17 per cent of domestic production in 1972 to 50 per cent in 1976. It should be added that the world textiles industry was generally suffering from competition from newly industrializing countries such as South Korea and Taiwan, in the 1970s and it was precisely at this time of intense international competition that the textile industry in Iran also faced difficult domestic conditions. For woollen textiles, although production in 1977 was somewhat higher than the 1972 level, the volume of imports was even higher, reaching almost 95 per cent of domestic production although, due to the lack of data, it is difficult to estimate the volume of such imports for 1972. Comparing the domestic ex-factory price of a yarn of wool with the CIF import price, the 1979 ex-factory price was 73 per cent of the 1979 CIF price (Table A.22). Alternative evidence which is consistent with imports undermining domestic production is that excess capacity in the textile industry was found to be 61 per cent of total capacity in 1977.[10] In other words, imports penetration led to excess capacity initially and most probably to capacity elimination in the medium term.

Moreover, the number of workers in the large textile factories declined by almost 30 per cent in the same period (Figure A.4). Given the large share of IMDBI's loans to the textile sector, some increase in the mechanization of the sector appears to have also happened (Table A.23).[11]

If it is remembered that the textile industry in the 1960s was among the industries with low relative price, and cotton textiles in particular, had high value added, it becomes clear that adjustment to the oil boom has resulted in a squeeze of this important traded sector. Rising wage and material costs plus power shortages led to higher production costs, and competition from imports limited the ability of the industry to raise prices in spite of growing domestic demand. Higher costs and limited increases in price due to increasing international competition resulted in a double squeeze of the textile industry.

Footwear

Another manufacturing industry in the group of high domestic value added and relatively low price compared to the border price is footwear. Footwear production rose at 9 per cent per year during 1972-8. Footwear production relies on domestic inputs but production is highly capital-intensive which means that there were offsetting factors in this sector, making the outcome difficult to predict. In 1976, 85 per cent of footwear production came from automated and semi-automated plants (IMDBI 1976, p. 27). Also, Iran exported a considerable quantity of shoes to Eastern Europe in the early 1970s. In 1972, of the 80 million units of the total domestic production, 10 million units (12.5 per cent) were exported (IMDBI 1973, p.12).

By 1977, however, domestic production had reached 128 million units, and only 3.5 million units (2.7 per cent) were exported (Table A.21). The decline in exports is attributable to higher domestic demand as well as some erosion of competitiveness due to higher material costs. Empirically, it is very difficult to separate these two effects. In applying the adjustment model of Chapter1, therefore, one has to be careful not to attribute the decline in non-oil exports solely to the higher real exchange-rate (that is, price effects) and account must also be taken of the income effects and income elasticity of demand for various goods, as demonstrated by the example of shoes which have a high income elasticity of demand.

Food processing

The number of firms in food processing grew gradually until 1975–6, after which it began to decline. Although the number of firms declined in this sector, the physical volume of production of canned fruits and vegetables, and of vegetable oil, grew at the relatively fast rates of 21 and 10.8 per cent per year, respectively between 1972 and 1977. So the decline in the number of firms does not necessarily reflect the decline of the industry as a whole. Most probably, some degree of industrial concentration and greater mechanisation took place, especially since, from 1974 onwards, the food processing industry received a large share of IMDBI's loans (Table A.23).

It should be added that not all the food processing plants had an increase in the volume of production. The output of sugar factories reached a peak of 700,000 tonnes in 1974, after which it declined to 630,000 tonnes in 1977 which was lower than the 1972 level. For example, the high value added of the sugar industry (Table A.20) means that it must have suffered from the rapid increases in wages and material costs. Moreover, since the government was subsidising sugar consumers, the rise in the factory price was limited and it must not have been sufficient to compensate for rising costs.

Transport equipment and metal products

A similar trend in the number of firms – decline after 1975 – is visible for transport equipment and metal products. But again physical volume did not show a corresponding decline. Production of passenger cars, for example, rose from 49,000 units in 1972 to almost 140,000 units in 1977, implying a growth of 23.3 per cent per annum. From Table A.20 it can be seen that passenger cars are heavily based on imported inputs and they remained a heavily protected sector. In other words, industrial survival favoured particular industries – those based on imported inputs, those with high capital intensity, and those with high divergence between domestic and world price. This is an interesting finding since it shows that it is not so much 'deindustrialisation' but a kind of perverse 'reindustrialisation' which has resulted from the post-1973 adjustment policies followed in Iran, since a number of industries which are import-dependent, capital-intensive, and protected in fact grew. The reason for this perversity is that such a structure may not be viable when the oil boom is over and the oil doom begins. Such an industrial base is also likely to have input-supply problems with each turnaround in the price of oil.

The above discussion again demonstrates that a purely demand-based, relative price analysis, *à la* 'Dutch-Disease', is likely to ignore the complex interrelationship among the macroeconomic policy of high expenditure, the nature of intermediate inputs in industrial production, and sectoral policies of the government, including protection and credit, which together determine industrial development and supply response in various industries. The nature of intermediate inputs, commercial policy, and credit policy can prove to be important offsetting factors in lowering the adverse impact of real appreciation on the traded manufacturing sector (Chapter 1).

Non-metal mining products

Another impressive feature of Chart A.3 is the phenomenal growth in the number of firms classified as processing, non-metal mining products, the most important of which are glass, bricks and cement – those, in other words, mostly related to construction activities. The production index of the construction-related industries rose during the 1970s on average by 13 per cent per year. Bricks, for example, are non-traded due to their relative high bulk and low value, therefore, brick producers did not face competition from abroad and domestic prices rose considerably. Brick production grew rapidly at 17.4 per cent per year between 1972 and 1977 (Table A.21). Brick-making in the 1970s was largely a labour-intensive rural industry.[12] Brick making, however, was largely staffed by Afghan workers in this period and this contributed to lower wage costs (Madjd 1983, p. 24). The presence of foreign workers also raises a broader issue – how labour can become a traded input. This again is a policy issue which depends upon the government's attitude towards migrant labour.

Cement production also showed a rapid growth. Cement production has high value added and it was priced considerably below the world price both in 1970 and in 1979 (Table A.21). In spite of a rapid rise in production and some imports, there was a considerable cement shortage, principally based on growth in domestic demand for a largely non-traded input. Almost half the total supply of cement was destined for military projects which crowded out civil construction works. Although official prices remained stable, black-market cement prices were five times the official price.[13] Therefore, the growth of cement production was due to the highly mechanised nature of the industry and high black-market prices. Moreover, IMDBI had considerable investment committed to cement production (IMDBI 1976, p. 59). It should be added that the Pahlavi foundation (operated by the Shah and his family) owned substantial shares in cement factories around the country and getting a permit for establishing a new factory required the consent of the Pahlavi foundation. The author is aware of an interesting case concerning the refusal of the authorities to allow the establishment of a new cement factory at a time when cement was in severe shortage. The application was for a new cement plant in the Damghan region where all the raw materials and sufficient labour were available (early 1975). The machinery for the plant was also available to the applicant at concessional terms.

The Ministry of Industry, however, refused the application on the grounds that such a factory would not be able to sell its production in the Damghan area since the area does not have the capacity to absorb the proposed volume of production. In fact, Damghan is only 400 km from Tehran, which was a key market for cement, and other existing plants were much further away – for example, the plant in Isphahan is 1,000 km from Tehran, that in Kerman is 1,600 km away, and that in Doroud is 800 km from Tehran. Moreover, in 1975 large quantities of cement were being imported from abroad. The justification for refusing the application was clearly insufficient. Private discussions with the Minister in charge revealed that he had no say in granting the application. It became evident that the authorities were not interested in increasing the value of domestic production. This is another example of the key importance of considering policy factors in explaining supply response (or lack of it).

The evidence derived from the analysis of the manufacturing sector, therefore, reveals how both economic forces and policy decisions interact to bring about a particular supply response which is not predictable from a purely economic model of the 'automatic adjustment mechanism'. It has been shown how the rapid industrial growth of Iran in the 1960s and early 1970s came under severe pressure after 1975 as a result of economic policies after the oil boom. Rising prices and increasing scarcity of non-traded goods (for example wage costs, power shortages, and rising land prices) all contributed to a rapid rise in the cost of production. When industries received protection or when the product was non-traded, as in the case of bricks, higher domestic prices could compensate to some extent for rising costs. Also, higher capital intensity, such as in shoe-making, contributed to maintaining profitability, suitable for industrial diversification and viable after the oil boom, such as textiles, were squeezed because of higher costs and considerable competition from imported textiles, and as a result of policy factors. Industries with low domestic value added and high protection, for example private motor cars, fared better.

Summary

Looking at the performance of the manufacturing sector as a whole, we find three types of industrial performance in this period: first, declining industries; second, industries with slow growth (below 10 per cent per year); and, third, industries with fast growth (over 10 per cent per year). Table A.21 shows the average growth of these three types of industry. Although no absolute relation can be established, it is possible to observe a broad relationship between relative dependence on domestic resources (low import content), the level of protection, capital intensity, and output growth in the 1970s. Textiles and sugar are both in the high value-added category (Table 3.1) and they were among the declining industries.

The fast-growing industries include brick-making which is non-traded, passenger cars which was highly protected, and refrigerators which are also in the high price category. The other fast-growing industries are all in the highly mechanised group. In the slow-growth industries, footwear has high value added but production is highly mechanised, while televisions and sheet glass both enjoyed relatively high prices in international terms.

In sum, the macroeconomic framework of Chapter 1 provides the foundations for analysing the price changes affecting the manufacturing sector during the oil boom, both in terms of the relative price of traded and non-traded goods and in terms of wage–price relationship. Wages grew faster than the price of manufactured goods, as expected from the discussion of the theoretical model and the implied magnification effects. Nevertheless, this chapter has shown the importance of analysing other factors such as capital intensity, import dependence, price protection, and government credit policy, which can either offset or accelerate the adverse impact of real appreciation. The prices of traded goods fall relatively, and there is a general boost in domestic demand which means greater demand for domestically produced tradables in some instances, such as in home appliances or passenger cars. The price effect is not allowed to operate due to government tariff or import controls resulting in the expansion, and not the contraction, of domestic industry. Nevertheless, it has been necessary to complement the analysis by a detailed discussion of the government's tariff pricing policy concerning electricity, and some broad indication of its credit policy through the IMDBI in order to explain supply response.

The role of credit has not been substantially illuminated in this analysis due to the difficulties of obtaining data on the subject. Broadly, it should be added that, in this period, imported capital goods become cheap relative to labour and industries engage in greater use of imported machinery. But the degree of actual substitution would depend, among other things, on the degree of the firm's access to credit which in turn is very much influenced by the government credit policy.

It is worth noting that the IMDBI's financial assistance to the private sector was confined to a small group of manufacturing enterprises which were the largest modern manufacturing enterprises in the private sector. According to Alizadeh (1985), between 1959 and 1978, IMDBI's total loans of 137.5 billion rials (nearly $2 billion) to manufacturing industries was channelled into only 650 enterprises. These enterprises were among the 926 largest manufacturing establishments. In other words, the bulk of industrial credit was largely out of the reach of smaller enterprises, largely based on the bazaar for their marketing, which are largely dependent on the local economy for their inputs.

This is especially important since it is likely that commercial credit would shift largely towards the non-traded goods, leaving the traded sector very much dependent on official sources. This point further highlights the importance of analysing sectoral-level policies for explaining supply response in the adjustment process. To repeat, although the macro adjustment model is powerful in explaining demand changes, it clearly is insufficient to explain supply response.

Finally, it has been shown how short-term demand factors overrode non-oil based supply advantages through eroding profitability from a number of industries which were previously viable. The set of policies both at the macro and at the sectoral level were, therefore, detrimental to the development of industries which do not depend on large oil earnings or on the demand forces generated from the spending of the oil revenue. In summary, for the period under discussion

1. The Dutch-Disease model presented earlier does explain to a considerable extent relative price changes facing the manufacturing sector.
2. Supply response is not adequately explained by the Dutch-Disease model.
3. The policies followed by the government at the macro level have generated an adverse set of relative prices facing the manufacturing sector.
4. Industrial policies at the sectoral level have largely benefited industries which produce non-traded goods, industries which are import-dependent and mechanised, and in many cases those which are largely protected.
5. Macro policy choices and sectoral policies towards the manufacturing sector have resulted in the decline of those industries with greater reliance on domestic resources and less protection and have made the manufacturing sector more vulnerable to changes in oil earnings.
6. An alternative set of macro and sectoral policies is theoretically possible with less adverse effects on industries which rely on renewable domestic resources, on industries which are less dependent on protection, and export-orientated industries, hence reducing the economy's dependence on oil. Also, small industries could be given greater support since they are more reliant on local initiatives and needs, less dependent on foreign exchange and imported technology, and provide a better opportunity for a broader spread of the benefits of industrialisation in terms of employment. Although such policies may not be agreeable to the contemporary elite in power, nevertheless, these are likely to result in generating a more viable economic and political base for the potential state which would implement them.

Notes

1. The figures exclude oil but include mining, which has never been more than 5 to 6 per cent of the total value added in manufacturing and mining so the growth is mostly due to the growth in manufacturing activity. For convenience, in the rest of the chapter we shall refer to manufacturing and mining simply as manufacturing.
2. Calculated from Tables A.8 and A.9.
3. The following discussion on commercial policy and financial incentives relies primarily on Sadigh (1975).
4. The following discussion on import dependence relies mostly on Avramovic (1970).
5. Classification of value added is in relation to dependence on imported inputs. 'High value added' are generally those industries where the dependence on imported inputs is below 33 per cent of the value of output (sales price); 'Low value added', where the dependence is above 33 per cent. Classification of prices is in relation to import prices CIF, net of duty.

'Relatively low' prices are generally those where the excess of the domestic price over the import CIF price is below 33 per cent; 'Relatively high price', where the excess is above 33 per cent. For the price level the 33 per cent figure was derived from a simple ranking of industries according to the level of the excess of domestic prices over CIF import prices, which yielded a median value of 33 per cent with a range of between 0 and 200 per cent. For value added the 33 per cent was derived from analysing 37 product lines in less than half of which (16) the share of imported inputs was less than 33 per cent.

6. Conceptualising Table 3.1 in terms of effective protective rates, we can say that in categories A and C high value added or low import content means that inputs are are not subject to tariff, and therefore category A would have lower effective protection than category C. In categories B and D low value added or high import content means that there is likely to be an import tax on intermediate inputs, and therefore category B has lower effective protection than category D. The comparison of A and D or of B and C is more difficult as far as effective protection is concerned since it is not clear to what extent the tax on intermediate inputs offsets price protection.
7. The following discussion on electricity draws from Shafaeddin (1980, pp. 127–34).
8. Calculated from the figures provided by the Ministry of Power, *Sanate Bargh*, 1976, p. 1, cited in ibid., p. 133.
9. The author himself remembers that, because of price controls on ice in summer on 1975, no ice could be found in any of the three large ice factories visited on the Caspian coast and this was at the peak of the tourist season. One of the factory owners explained that the official ice price was far below the cost of production.
10. US Embassy, *Semi-Annual Economic Trends Report*, Tehran, May 1977, cited in Graham (1979, p. 120).
11. There are a number of measurement problems which inhibit the use of a more direct indicator of capital intensity such as the incremental capital/output ratio. First, as has already been mentioned, there was evidence of widespread excess capacity in the industrial sector of Iran in this period which makes the use of the capital/output ratio unreliable. Second, even if there was no excess capacity, the absence of an industry-level price deflator for investment and for output further reduces the reliability of incremental capital/output ratio comparisons.
12. By 1976, only 13 per cent of bricks were machine-made (IMDBI 1976, p. 29).
13. In 1977, the official price was 200 rials for a 50 kg bag, while contractors were paying up to 1000 rials (Graham, 1979, p. 107). It should be added that there was also considerable military use of cement which contributed to the shortage.

4 Agricultural production in Iran

This chapter is a detailed analysis of the agricultural sector in Iran during the 1970s. The questions addressed in the chapter emerge directly from the theoretical framework presented in Chapter 1. These relate to the analysis of the relative price developments of the traded and non-traded agricultural products, input utilisation, the supply response of the sub-sectors producing these crops, and a consideration of how government policy has affected the outcome. Moreover, the chapter attempts to explain the supply response by analysing the agrarian structure and other institutional factors affecting supply response. A comparison is made between the industrial and agricultural sector in terms of the response to the adjustment process and relative price changes. The comparison reveals a fundamental contrast between the supply response of traded goods in the two sectors. It is argued that this is the direct result of pro-industrial bias of government policy and, more specifically, a result of the greater access to credit in the industrial sector in general and selected industries in particular. In other words, although capital became more abundant relative to labour and other localised non-traded goods, its distribution remained highly skewed, disadvantaging the majority of agricultural producers.

The performance of the agricultural sector in the 1970s requires not only an analysis of the relative price changes as they affect the output and input price, but such relative price changes must be placed within the structural context of the rural areas. This is done in terms of both the agrarian structure and the allocation of labour between agricultural and non-agricultural activities in the rural areas. Indeed, as we shall see, one cannot adequately explain the response of agricultural output and its composition solely in terms of the Dutch-Disease adjustment model. This is because a number of institutional factors, such as the nature of the land market and the fragmentation of agricultural land-holdings which resulted from the land reform carried out in the 1960s and early 1970s, and the nature of the capital market and the access to credit, are also important since they affect response and adjustment in the Dutch-Disease model. Moreover, as has been emphasised throughout this book, the analysis of government policies towards agriculture is also essential for an adequate explanation of the supply response.

The chapter begins with a discussion of the land reform and its impact on the agrarian structure. This is followed by a discussion of the government measures designed for modernisation of agriculture which accompanied and followed the land reform. Next, we discuss the cost of production in the agricultural sector through an examination of changes in the opportunity cost of labour and its increasing scarcity for crop production. In this section, the main traded and non-traded crops are discussed and the differential impact of relative price changes of outputs and inputs on these different sub-sectors is analysed. The chapter concludes with a summary of the findings and notes an important contrast between the supply response of traded goods in the agricultural and the industrial sectors of the Iranian economy in the 1970s. The comparison is an attempt to link the analysis of the changes in the agrarian structure with output response. It is argued that the land reform and the government's pro-industrial policy generally reduced the creditworthiness of the bulk of the agricultural sector. This reduced the ability of that sector to compete for resources and technology with the consequence of a more uniform adverse impact of the adjustment process on that sector than on the rest of the economy.

Therefore, before entering into the more detailed analysis of the price changes of major crops, the cost of production, and output response, it is important to consider the impact of the land reform on the farm structure and on the distribution of agricultural resources such as land, water and credit. The supply response of the agricultural sector, therefore, will be located not only within the Dutch-Disease framework but also within the broader socio-economic and policy context affecting that sector.

Land reform and farm structure

The land reform which took place in the decade prior to the oil boom consisted of land purchases from the owners and its subsequent transfer to the cultivating peasants and sharecroppers. This transfer involved radical changes in the pattern of ownership and resource use in Iranian agriculture. There were several major impacts of the land reform which have direct bearing on agricultural performance in the 1970s.

First, the transformation of traditional village-collective farming to small-scale owner-cultivator family farming increased land fragmentation. This lowered the mechanisation potential, thus lowering the ability of that sector to respond to rising wages, resulting in lower output.

Second, the land reform contributed to the disappearance of the *boneh*. This was a co-operative form of water use highly suited to farming in a situation of water scarcity; in addition, the *boneh* prevented excessive fragmentation of the fields which would have been the result of the Islamic inheritance laws. Hence, the passing away of *boneh* was a contributory factor in bringing about land fragmentation.

Third, the land reform brought about a high degree of uncertainty concerning property rights over the land, and this was an important factor behind the inability of agriculture to attract sufficient amounts of capital and commercial credit. The absence of appropriate deeds concerning ownership of agricultural land also hindered farm consolidation, further hindering the access of small holders to credit. In other words, an important institutional factor – the land reform – must also be considered before we can appropriately place the relative price movements which we shall describe in the next section. This is an important point since it implies that the Dutch-Disease model (the macro-economic adjustment model discussed in Chapter 1) can only be used as an explanatory framework if it is appropriately integrated into a broader appreciation of institutional and long-term structural changes of an economy. To consider the above factors in more detail, we now proceed to describe the agrarian situation before and after the land reform.

Land-ownership prior to the land reform

Prior to the land reform, there were four broad categories of land-ownership. These consisted of the public domain lands (*khalaiseh*), crown lands (*amlaak*), lands endowed for religious and public purposes (*waqf*), and private holdings (*melki*). The quantitative extent of each category, however, is subject to dispute since land registration was hardly developed in many places. Nevertheless, according to Dehbod (1963), of the 51,300 villages recorded in the 1956 census, the structure of land-holdings was as follows: 10 per cent public lands, 4 per cent crown lands, 10 per cent religious endowments, 76 per cent private lands.

The land-tenure system, on the other hand, was placed under three categories: large-scale share-cropping (*mozare'eh*), family-size tenancies (*ejareh'ee*) and family smallholdings or owner-cultivators (*melki*). These tenure arrangements were common to all the different forms of ownership regardless of their status as *khalaiseh, amlaak, waqf, or melki*. The extent of these different forms of tenure is also difficult to assess. A survey by the Statistical Centre of Iran gives the following figures concerning the relative importance of each type of tenure prior to 1962: 44 per cent of units and 55 per cent of area was under sharecropping, 12 per cent of units and 7 per cent of area under tenancy, 33 per cent of units and 26 per cent of area under owner-cultivators, and 11 per cent of units and 12 per cent of area under a mixed form (Ashraf and Banuazizi 1980). Large land-holdings organised through share-cropping, also known as the system of arbab–rayati (master – subject), was therefore the most prevailing form of agricultural organisation prior to the land reform in most regions, with perhaps some exceptions such as the Caspian lowlands. In the Caspian littoral and parts of western Iran where water is plentiful it was not sharecropping but tenancy which was the dominant form of land-tenure (Ashraf and Banuazizi 1980).

Moreover, in regions of water scarcity, sharecroppers would normally cultivate the land in farming groups known as *boneh*.[1] Each village in such areas would be divided into several *boneh* each cultivating a given area of land with a specific number of peasants. The number of *boneh* in a village was determined by the availability and frequency of irrigation water, which in most instances was provided by the *ghanaat*. *Ghanaat* are artificial underground water channels. Their function is to conduct underground water to the surface some distance away. By appropriate design, the *ghanaat* water can surface near the village crops. *Ghannat* are expensive and difficult to construct and require constant maintenance in order to keep them open. They were mostly maintained and extended by the members of the *boneh* themselves although the initial capital cost of construction and repair would in most instances be born by the landlords who would levy charges upon the harvest of the *boneh* (Hooglund 1982, p. 105). As a social institution, the *boneh* was a medium for solving conflicts among the peasantry. The head of each *boneh* (the *sarboneh*) would arbitrate peasants' disputes in many areas including the allocation of irrigation water.

Now let us consider the land reform itself.

The land reform, 1962–72

The reform was carried out in three phases and was officially completed in 1972–3. Each subsequent phase may be viewed as an attempt to respond to the shortcomings as well as the impact of the previous phase. In the first phase of the reform, the land-owners could no longer own more than six *dong* (a *dong* being a sixth of a village), that is, a village or its equivalent scattered in different villages. The land-owner was allowed to choose any six *dong* of his lands but had to sell the rest of his estates to the government, which then transferred the property rights to the occupant sharecroppers. Exceptions to the first phase included mechanised lands, public *waqf*, orchards and plantations. The number of the so-called mechanised units was relatively small. In 1960 it was estimated that about 10 per cent of all agricultural units used mechanical ploughing, of which 95 per cent relied on tractor rentals and custom ploughing. Also, most of the stock of agricultural machinery was located in the north and north-eastern provinces of Gorgan and Mazandaran (Nowshirvani 1978, pp. 27-8 and 33). Concerning orchards, groves, vineyards and plantations, the land-reform law declared the land to be the property of the landlord. Nevertheless, roots, plants and trees became the property of the peasants. This meant that although sharecroppers retained possession of the land, they could not legally sell the property without the consent of the landlord.

The land was distributed not in plots of a given size but according to the existing *nasaq*, or customary right to occupy land. The *nasaq* gave the *nasaq*-holder the right to cultivate a portion of an owner's land and use village water. A *nasaq*-holder, however, would not work an assigned plot of land alone but by belonging to a *boneh*. In villages affected by the land reform each *nasaq*-holder acquired title to the land on which he was working at the moment. In many cases the land would be assigned to a *boneh*, rather than to an individual *nasaq*-holder, and the members would divide the *boneh* field among themselves. Given the division of *boneh* land into excellent, medium, and poor, the equal distribution of the *boneh* land implied that each member would receive several parcels of land from each category.

Peasants receiving the land were required to pay the purchase price in 15 years plus 10 per cent interest. Membership in a co-operative largely designed for credit was also a condition for receiving land. Moreover, the local co-operative society would become responsible for the maintenance of irrigation facilities, normally beyond the means of the individual peasant.

There are varying estimates concerning the actual number of families affected under phase one although The Ministry of Rural Affairs and Co-operatives stated that 16,246 villages were distributed between 781,322 families under this phase.[2]

The first phase of the land reform encountered a number of problems. First, many peasants who were working in the 'choosen' villages of the land-owners simply did not receive any land. In the 'chosen' villages, the landlords continued to cultivate the land on the old sharecropping basis. Thus, the peasants of one village might have lands transferred to them, whereas land had not been transferred to their neighbour (Lambton 1969, pp. 112–30). Second, no limits had been set for the maximum size of the exempted areas, so that large areas of land had remained in the hands of the landlords due to their status as 'mechanised'. This created sharp criticisms and was seen as a betrayal of the original purpose of the reforms, leading to reduced confidence in the seriousness of the reforms. Third, the countryside had become chaotic, especially since many peasants working in the chosen villages had refused to pay the landlord's share that year. Finally, smaller land-owners were placed under uncertainty due to the possibility of the extension of the reforms to their villages. The result had been the reduction of the volume of investment in agriculture, leading to the decay of such important infrastructural facilities as the underground water channels or *ghanaat*.

By January 1963, new amendments were added to the 1962 law. The landlords were now given one of several options: renting the land to the cultivators for 30 years, or 99 years in case of the land endowed for religious purposes, and the occupying tenant would pay a fixed rent based on the average income received by the land-owners in the preceding three years; selling the land to the cultivator directly or to the government, much as in phase one; dividing the land with cultivators according to the existing crop shares, and setting up joint-stock companies with the cultivators. By 1966, the options chosen were tenancy 74.63 per cent; division 10.61 per cent; joint enterprise; 10.60 per cent; sales 3.18 per cent; and tenant right sale 0.98 per cent (Denman 1973, pp.136–7). In phase two most land-owners had refused to sell the land to the peasants; hence tenancy was the next best alternative for both parties, followed by sale or joint enterprise. The beneficiaries of phase two were over 1.6 million peasant families (Madjd 1983, p. 9).

As phase two approached its completion, both the tenants of the 30-year and 99-year leases had become radicalised to different degrees since they had not become the owners of the land they cultivated. On occasion they would band together and withhold rent (Afshar 1972, cited in Denman 1973, p.144). Rent was also withheld in some of the *waqf l*ands, and there are instance of takeover of some of the non-cultivated areas by tenants (Mo'meny 1980, p. 241). The general instability in the rural areas resulted in the introduction of yet another phase in the land reform; in fact the need to maintain stability and calm in the rural areas was given by the Minister of Agriculture as the main reason for this stage (Mo'meny 1980, p. 239). It should be added that the appearance of new opportunities of investment in the non-agricultural sector of the economy had created a desire for the smaller land-owners, whose lands were frozen under the 30-year leases, to sell the land and invest in other areas (Mo'meny 1980, p. 240).

Phase three of the land reform was introduced in 1968 in the 'Bill for the Sale and Distribution of Leased Lands'. According to this law, the landlords were required either to sell or to divide the land. By 1973, the title of 282,000 estates were transfered to 752,600 households and 90,000 estates were divided (Ministry of Co-operation 1976, cited in Ashraf and Banuazizi 1980, p. 33).

Altogether, in the three phases of the reform, more than 3 million producers were affected (Amuzegar 1977, p. 221; see also World Bank 1975, cited in Madjd 1983, p.10). An indication of the number affected is also given by the co-operative membership figures, since joining a co-operative was a precondition for receiving land. The first column of Table A.27 shows that by 1972–3, the year of the completion of the land reform, 2.065 million farmers belonged to the co-operatives. Also, between 1972–3 and 1977–8 co-operative membership expanded to 3.001 million members, which gives credence to the figure of 3 million producers being affected by the reform.

The structure of landholdings and output shares for 1972 has been estimated in a World Bank study (Price 1975) using a 1972–3 survey of the Iran Statistical Survey. The results are shown in Table A.26. The farm structure consisted of 2.533 million privately owned farms with an average size of 6 ha.

There are 7,000 farms larger than 100 ha, with an average size of 258 ha and constituting 12 per cent of total farm area. In the next category there are 10,000 farms with 51–100 ha, with an average size of 70 ha. Next are the 11–15 ha farms, which are the most important in terms of land area, occupying 46 per cent of total farm land, with an average of 12.5 ha and constituting 16 per cent of the farming families. Last, we come to small ownership holdings of less than 10 ha, which constitute 37 per cent of farming area and consist of 84 per cent of the total peasantry. It may be noted that farms of less than 1 ha were occupied by 32 per cent of all farming families. Putting the result together we have 0.03 per cent of the families had 12 per cent of the land, 15.9 per cent of the families had 50 per cent of the land, and 84 per cent of the families had the remaining 38 per cent of the land.

Madjd (1983, p.14) however, has argued that the survey does not reflect the final results of phases two and three of the land reform, especially since the number of farms larger than 51 ha is exactly the same as the number of landlords – 17,000 – who opted for the division of the land which was to be further subdivided. In other words, it is likely that the farm structure created by the land reform was much less concentrated than indicated in Table A.26.

Concerning the shares of output, the land share of largest farms of over 100 ha was 12 per cent, but they only produced 6 per cent of the total output. The medium-sized farms produced 36 per cent of the output from 50 per cent of the area. The small farmers produced 41 per cent of the total output from 36 per cent of the total land. The remaining 17 per cent of the output was produced by 'pastoralists' and other rural people not owning land. In terms of the marketed surplus, however, the situation is quite different. The contribution of small farms was only 19 per cent and the rest was produced by the medium-sized and large farms. In fact, the latter group produced most of the grains, such as wheat and barley, whereas the small farmers would concentrate on labour-intensive cash crops such as fresh fruits and vegetables, tea, tobacco and rice in the Caspian area.

These figures indicate the prevalence of small farms among the total number of farmers but a fairly unequal distribution of the land area between large and small farms, with a Gini coefficient of about 0.62 (calculated from Table A.26). Taking into account the later division of the 17,000 farms classified as large farms, more than 84 per cent of peasant families own less than 10 ha and more than 32 per cent have below 1 ha. It should be added that in most instances the size does not refer to a single piece of land but to several plots scattered throughout the fields. On average 7 ha (of which half remains fallow) with a yield of 1.3 tonnes of wheat per hectare is accepted by Iranian agronomists as the minimum requirement for subsistence for a family of six (Hooglund 1982, p. 77). Given that over 67 per cent of all peasants acquired less than 7 ha, the size of land obtained by most peasants must be considered as below the minimum necessary for subsistence. In any given year, only half of the land can be cultivated while the other half is left fallow because of water shortage, In effect, the land reform created a small class of medium sized commercial farmers and a mass of poor peasants.

It is clear that survival of such a large stratum of the peasantry requires income from outside employment. Both informal urban and off-farm rural employment were important sources of outside income for the marginalised peasants. The analysis so far demonstrates the importance of analysing the agrarian structure for understanding the nature of the labour market and the supply of labour and its response to both rural and urban income opportunities.

One important factor affecting the labour supply in the rural areas, which further illustrates the forces behind the migration into the cities during the 1970s, has been the way land reform based itself upon the existing differentiation of the peasantry. The focus of the reform was on those sharecroppers who owned one or more farming assets such as oxen, and comprised the better-off section of the sharecroppers. Such a cultivator, legally referred to as a *zaareh'*, received top priority in the distribution of land. After the *zaareh'*, next highest priority was accorded to their descendants if the original sharecroppers had died within a year of the land reform. Third in line were the sharecroppers referred to as the *barzegaraan*, who, unlike the *zaareh'*, possessed no farming assets apart from their labour and the traditional right to use the land (*nasaq*). The lowest priority was accorded to the farm workers proper, or the *kaargaraane keshavarzi*, who were not sharecroppers with cultivation rights (*nasaq*), but day labourers who normally worked in different lands at different times. Very few of these day labourers, if any, received land.

The day labourers are normally classed with the rest of the landless villagers *khwushneshin* and they constituted a significant proportion of the total population in each village. Officially, it was argued that with the establishment of family farms, the need for day labourers in the village would be reduced, hence they could be transferred to other sectors of the economy. According to the surveys of 1960, 1962, and 1965, the *khwushneshin* population comprised 40-50 percent of all village inhabitants (Hooglund 1973, p. 230). Assuming 5 per cent of the *khwushneshin* population to be in occupations other than day labouring, something like 1 million families of agricultural labourers were excluded from land redistribution (Khosravi 1976, p. 130).[4] This, coupled with some mechanisation among the medium-sized farmers, resulted in massive unemployment among the *khwushneshin* population, hence many of them, particularly the younger ones, migrate to the cities.

Another aspect of the land reform was the disappearance of the *boneh* which, as was mentioned above, was an important institution for resolving conflicts, particularly with regards to the allocation of water. Prior to the reform, membership of the *boneh* also ensured a share of the available water to the cultivating peasants. The land reform redistributed most plots along with the rights to a fixed amount of water at regular intervals. Where the source of water is an open stream or a river, traditional rights have been observed. Nevertheless, disputes have been common in villages which depended upon the *ghanaat* for most of their water requirements (Hooglund 1982, p. 91). It has been common for landlords to keep ownership of *ghanaat* so as to ensure adequate water for their fields and orchards as well as to sell the excess water to the local farmers. One study of the Kerman province documents that since 1971 absentee *ghanaat* owners have demanded high rents for the use of *ghanaat* water. Another problem related to irrigation includes changes in the flow of *ghanaat* water due to the introduction of new crops with higher water requirement in land kept by the owners. Furthermore, although some *ghanaats* fell under village control, the introduction of deep wells by commercial farmers has substantially reduced the availability of underground water for the smaller farmers.

Another impact of the land reform was the creation of a highly distorted land market. This distortion came about since the recipients of the farm lands were barred from the sale of the farm lands until the completion of payments, which in most cases would last between 15–20 years. This meant that the vast majority of the farmers were without legal deeds establishing ownership. This created a dual market for land with those lands with proper papers fetching a premium price. This situation in the land market also prevented farm consolidation and mechanisation, thus lowering the potential output of agriculture. Furthermore, the uncertainty concerning legal titles was an important factor behind the failure of agriculture to attract sufficient capital and commercial credit. The uncertainty over property rights plus, as we shall see, a government policy which systematically discriminated against small farms in terms of resources resulted in lowering the output potential of small farms which are the majority of the farmers in Iran and thus also the aggregate supply response of the agricultural sector.

Before concluding this account of the land reform and discussion of the performance of the agricultural sector in the 1970s, it is also necessary to analyse the attempts made by the land reform laws to modernise the traditional farming structure to deal with the problems which arise from the fragmentation of holdings after redistribution.

Modernisation measures during and after the land reform

In this regard, measures included the creation of two main types of organisation, namely rural co-operatives and farm corporations.[5]

Reference has already been made to rural co-operatives, the membership of which was in most cases compulsory before the peasants could receive land. It was hoped that through these co-operatives the credit and some of the services which the landlord had previously supplied could be provided. Also, the co-operatives were expected to reduce the dependence of the peasants on the rural money-lenders. Table A.27 provides a general picture of the performance of the co-operatives from 1963 to 1977.

Membership of co-operatives grew steadily to over 3 million and on the eve of the revolution nearly 85 per cent of the villages were covered. The total amount of loans provided by the co-operatives rose from 4 billion rials to 670,000 loan recipients in 1967, to 30 million rials going to 1.32 million loan recipients in 1977. The average loan increased from 6,080 rials ($86.20) in 1967 to 22,800 rials ($323.40) in 1977. Considering that the wage of a simple day labourer in the rural areas was 388.6 rials per day in 1977 (Statistical Yearbook of Iran 1980, p. 697), the average loan was equivalent to eight weeks pay for a day labourer.

Given the small amount loaned, it is likely to have been used mostly for meeting short-term consumption requirements rather than providing for capital expenditure essential for increasing productivity. In fact, the co-operatives were not even adequate for short-term needs. According to a 1968 survey, money-lenders accounted for 40–60 per cent of the loans received by the peasants and, given the smallness of the size of loans granted by the co-operatives, it is unlikely that the situation had improved during the 1970s (Gorgani 1971, p. 385). Interest rates charged by the money-lenders could reach as high as 200 per cent but on average, according to Gorgani (1971), they fluctuated between 24 and 120 per cent. Others have recorded an average of 30–50 per cent (Khosravi 1976). Given the average rate of inflation of around 15 per cent in the 1970s, the real rates are considerable. Sometimes the peasants had to resort to pre-selling the crop which could lead to further indebtedness or even ruin if they had to sell the whole crop in order just to meet the loan payments (Gorgani 1971, p. 384).

The other government credit institution was the Agricultural Development Bank of Iran (ADBI), which concentrated on large capital-intensive projects with a participation rate of nearly 50 per cent in the total investment project. The amount of credit and direct aid approved by the bank rose from 1.3 billion rials in 1971 to a peak of 48.6 billion rials in 1975, falling to 25.8 billion rials in 1976 (Table A.28). The average size of loan and the number of recipients contrasts sharply with that of the co-operatives. The average ABDI loan in 1971 was 18.8 million rials, compared with 8,015 rials provided by the Co-operatives. The number of loans approved in 1971 by ADBI was only 69, whereas the co-operatives gave loans to 876,000 farmers. The corresponding figures for 1976 can be seen in Table A.27; they do nothing to alter the picture. The concentration of ADBI loans and the massive size of each loan effectively meant that only the most powerful and richest farmers could have access to the loans. Included in the ADBI loan figures are the amounts paid out to unsuccessful agribusiness ventures set up with foreign capital participation. These are discussed later in the chapter.

Another measure adopted by the land-reform law to modernise the traditional farming structure was the establishment of farm corporations. This was an integral part of phase three of the land reform and its aims were stated as the increase in both output and income in agriculture. The idea was the consolidation of the peasants' scattered plots into large farms, along with scientific mechanised farming. The government had decided to subsidise these corporations for the first five years in order to help them to become viable business enterprises.

The peasants who had surrendered their lands would become stock-holders of the corporation which would be managed by a state official, preferably an agricultural engineer. They were to be established either in areas where 51 per cent or more of the inhabitants would agree to participate, or in villages which had been recently subjected to natural calamity. Moreover, the corporations were not conceived only as production units, but as more general social units. The stock-holders were to live together in 'modern villages', equipped with a school, a consumer co-operative, a house of culture, and so on.

Fourteen farm corporations were established in 1968. By 1976, the number of farm corporations had risen to 85, covering 778 villages and farms over an area of 310,000 ha. They had 32,500 share-holders and almost $20 million worth of capital (Ministry of Co-operation 1976, p. 38; Ashraf and Banuazizi 1980). In 1974, the gross income of 53 corporations amounted to $41 million, their book costs ran at $26.7 million, resulting in an accounting profit of $14.3 million dollars. The average annual profit of the 18,605 share-holders was about $801, the average annual wage of the share-holders who received employment in the corporation amounted to $400. It must be added that, from 1968 to 1974, the government had granted $68 million in free financial assistance to 65 corporations (Ministry of Co-operation 1976, pp. 79, 82 and 84; Ashraf and Banuaziz 1980, p. 41).

Official statistics aside, from studies of these corporations they appear to be something of a disaster. To begin with, they were not generally established by the will of a 51 per cent majority, but rather by the will of the Ministry of Agriculture wherever it saw fit (Mo'meny 1980, p. 363). Usually the farm corporations were established in some of the best lands located near the urban centres. These lands, contrary to the official claims, were largely expropriated from the peasantry by force, intimidation, or false promises (Mo'meny, 1980, p. 364).

The peasants were normally dissatisfied with performance of the corporation and many of them would have preferred to cultivate the land themselves. There were several reasons behind their dissatisfaction and even anger. First, their net income had been reduced (Salmanzadeh 1981, p. 215; Mo'meny 1980, p. 369). This was partly due to the heavy capital costs turning the corporations into loss-making enterprises; without the state subsidies, the stock-holders would not have been received the amounts they did. Furthermore, in spite of the promises made that the peasants could work for the corporations if they so chose, due to the heavily mechanised form of farming many of the peasants had become unemployed (Salmanzadeh 1981, p. 199). Also, the peasants were forbidden to keep flocks at their new 'homes'. This had resulted in the elimination of an important complement to the previous income of the peasants.

The second factor involved in the poor performance of the corporations and the dissatisfaction of the peasants had been poor management. The state manager would have little incentive for efficient production. He would receive a handsome salary every month regardless of the performance of the corporations. This lack of incentive, combined with a certain disdain for the peasant's way of doing things, had generated antagonistic relations between the stockholders and the manager.

Third, land-ownership for the peasants not only has economic significance but also social significance. The landed peasants have much more prestige in the village than the landless ones. In fact, becoming landless as a result of the takeover by the corporations was largely viewed as a return to their serf-like conditions. As Salmanzadeh (1981, p. 35) writes:

> 'To many of the peasants the manager represented a substitute for their former landlord and in this respect they petitioned him to intervene as an arbitrator in settling domestic family disputes. The manager's inability as a government employee to act in this paternalistic manner served to widen the distance between management and peasant share-holders'.

Farm corporations were basically disguised state farms. The alienation and the antagonism which exists between the peasants and the state in most societies was now concretised at the production level, resulting in the absence of any complementarity of interest between the producers and the managers. Such state farming also contradicted the original promise of the land reform, which was to give land to the peasants and not take it away from them. Lastly, from a purely technical, project-preparation point of view, they were badly designed and this aggravated the already unworkable institutional and organisational set-up.

Another of government's attempts towards agricultural modernisation was the establishment of agribusinesses for developing intensive farming on a strictly commercial basis. Agribusinesses were established in the late 1960s and early 1970s in areas which had become suitable for large-scale farming as a result of the newly created dams. Contrary to the measures for modernising agriculture, such as the farm corporations, these agribusinesses were not developed as an integral part of the land reform and were in fact quite inconsistent with it.[6]

On 27 May 1968, the law establishing 'Companies for Utilisation of Land Downstream of Dams' came into effect (Salmanzadeh 1981, p. 237). This law authorised the purchase of individual holdings in areas under the dams, for the purpose of consolidation and release to large agricultural companies for rapid development. They would be largely joint-venture operations with foreign capital. Each operating unit was to cover an area of 1,000 ha or over and the government undertook to provide for many of the needs of these enterprises, such as the primary and the secondary water channels and electric power. The investors were also exempt from paying customs duties on imported machinery (Salmanzadeh 1981, p. 237). To get an idea of the capital and the area of land under these agribusinesses, consider the four agribusinesses in the Dez Dam area which is located in the south-west part of the country. These were the HN Agro-industry of Iran and America, Iran-California Agribusiness Company, Shel-Cotts Agribusiness of Iran, and International Agribusiness Corporation of Iran. These had capital of $11.4, $3.3, $5.7, and $16 million respectively. The land allocated to each company was 20,627, 10,536, 14,736, and 16,860 ha respectively. Furthermore, the first three had 8,300, 4,540, and 4,256 ha respectively under summer and winter cultivation (Salmanzadeh 1981, p. 241).

Generally the agribusinesses performed poorly. The situation was so bad that after 1976 the foreign partners in three of the largest agro-industries withdrew and ADBI was obliged to take over the shares and active management of the concerns (Salmanzadeh 1981, p. 268, Madjd 1983, p. 45). The reasons behind their failure appear to be quite similar to what has already been said about the farm corporations, except that here the withdrawal of the foreign partners provides sharp evidence of poor performance.

Similar to the farm corporations, to provide land for the agribusinesses the newly landed peasantry was forcefully expropriated. The official claim was that the land was purchased at good prices, but in fact, most of the land taken over for this purpose was in the most fertile areas of the country and hardly any proprietor would have willingly surrendered such land for the new projects. This forceful expropriation had created great animosity between the peasants and the new companies.

Similar to the farm corporations, the agribusinesses and agro-industries were conceived as social units involving new resettlement centres. People from different villages taken over by the projects would be put together with little regard to the existing communal ties. Salmanzadeh (1981) quotes Duncan Mitchell's expression of 'open disintegrated communities' to describe the newly formed resettlement centres.

Along the same lines, in 1973, the government initiated a programme of establishing 'rural development poles'. That stated idea was to develop concentrated rural centres so as to facilitate the provision of technical and social services: 'it was expected that the country's 55,000 villages would eventually be merged into 4,000–5,000 large rural poles and resettlement centres' (Salmenzadeh 1981, p. 242). This would have meant the virtual elimination of all those villages which were not considered 'viable'. The intention of the government apparently was to increase the area operated by farm corporations, agribusiness, and the development poles so that they cover 25 per cent of total cultivated area by the end of the Fifth Plan in 1978 (Chavanian 1975). This project, however, had not spread widely when the revolution of 1979 occurred.

It is important to note that these later developments of opting for large-scale highly mechanised farming plus the resettlement centres typify the general reversal of policy which had occurred from the late 1960s. This reversal had implied a move away from the original idea of the land reform, that is, the establishment of small family farming, and a move towards larger-scale, American-style farming with the use of wage labour.

Let us summarise and further interpret some of the key points of the discussion concerning the institutional factors which have direct bearing on the supply response of agricultural output, before we analyse the relative price changes.

We have seen that the land reform created a small class of commercially orientated, medium-sized farms and a mass of poor peasants whose lands were insufficient for subsistence. This contributed significantly to migration to the cities, especially by the younger able-bodied sons. In addition, the land reform based itself on the existing differentiation of the peasantry; hence the better-off peasantry were treated preferentially while the poor peasants and the vast number of landless workers received no land. In other words, the land reform created a reserve army of labour which was to be intermittently absorbed mostly into the service and construction sectors in the 1970s. Part of this labour pool, the *khwushneshin* or the landless villagers, may be referred to as the 'surplus population' in Arthur Lewis's sense of the term, although seasonal employment in the rural areas would cast some doubt on this. But, the other part of this labour supply was drawn directly from the peasants whose lands were insufficient to provide their required income. This cannot be referred to as a surplus population since their transfer involved a net loss of output, although the contribution of such peasants to the marketed surplus was small.

Moreover, the land reform resulted in the distortion of the land market since the recipients of the farm lands were barred from the sale of the farmlands until the completion of payments, which in most cases would last 15–20 years. This meant that the vast majority of the farmers were without legal deeds establishing ownership. This created a dual market for land, raising the price of land with legal papers. The situation in the land market also prevented consolidation and mechanisation since land could not easily be transferred or sold. In addition, since the subsistence farmers could not sell the land, their migration did not usually result in the takeover of land by larger farmers and, hence, there was a net loss of output. In some cases sharecropping was re-established. The uncertainty concerning legal titles was an important factor behind the failure of agriculture to attract sufficient capital and commercial credit. This, and the fact that government policy systematically discriminated against small farms in terms of credit allocation, resulted in a further lowering of the output potential of the small farmers who are the majority of farmers in Iran.

In addition, there was a policy reversal away from small farms and towards large-scale, capital-intensive agriculture in the form of state farms and commercial agribusiness farms. This meant that the strategy would now ignore the difficult task of the progressive modernisation of the bulk of the nation's cultivators and concentrate its efforts on crash modernisation of the small sub-sector of large-scale mechanised farms. This 'bimodal' and skewed approach to agricultural development has been extensively criticised elsewhere (Johnston and Kilby 1975, Ch. 4; Lipton 1977). Aside from the equity and management aspects of such an approach, from a production point of view the quantities are likely to be insignificant in relation to the country's requirements. Also, such an approach is unlikely to be sustainable in the absence of large state foreign-exchange revenues to pump in operating capital. It would seem that, for an oil economy, the rational course to follow is one of maximum reliance on domestic factors of production, and especially on the mass of peasants who are willing to participate in the national effort of achieving a viable economy, develop their productivity, and reduce the country's heavy reliance on highly dependent technology and management.

Besides being generally unsuccessful, this rather arbitrary reversal of policy contributed to further uncertainty and created further disincentives for investment in agriculture, lowering potential agricultural output and increasing migration to the cities. The important point concerning migration to the cities is that there were a number of 'push' factors such as low rates of investment resulting from ill-defined property rights, peasants' indebtedness, and policy discrimination against small farms which should also be considered in analysing the movement of labour away from agriculture and towards urban-based activities; these affected the supply response of agriculture at both aggregate and crop level. This is not to deny that there are significant 'pull' factors, such as higher wages and better social services which further result in migration to the cities and drain agriculture of its labour pool, leading to stagnation or decline in output. These points and others are discussed further in the next section. Now that the institutional background is clarified, we proceed to investigate the impact of changing relative prices on supply responses. Policy issues are also brought in whenever they have a bearing on the argument.

Prices, costs and production after 1973

Now consider relative price movements after 1973. Let us look first at wage rates. Unfortunately, data specifically relating to the wage rates in the rural areas are not available. What is available is the opportunity cost of rural labour, namely the wages of the unskilled construction workers in the urban areas. This measure of opportunity cost is an important indicator, especially if it is compared with changes in agricultural prices. Together they provide an indication of the profitability of farming, although additional data on the rate of mechanisation would also be necessary for a full account.

Between 1973 and 1977, value added in agriculture rose on average by 2.5 per cent per year while the wages of unskilled construction workers more than tripled in real terms (see Tables A.9 and A.11). Wages rose despite the influx of workers into towns. Also, since the construction industry is seasonal because of the cold winter, the peak demand was at the same time as the peak demand for labour in the agricultural sector and this created further pressure on wages (Madjd 1983, p. 24).

The pressure on rural wages came not only from urban activities. Rural industries, such as brick-making and carpet-weaving, which are highly labour-intensive, were also expanding rapidly in the boom conditions after 1973–4.

Rising land values in the urban areas, and government regulations, meant that brick-making was increasingly located in the rural areas. The brick factories demand for labour was such that immigrant labour had to be used. Carpet-weaving experienced an unprecedented boom in the 1970s. A strong export demand combined with rapidly rising domestic demand led to high carpet prices. Carpet-weaving became a serious competitor for rural labour (Madjd 1983, p. 27).

According to a 1976 farm sample in the agricultural region of Neishaboor, of 104 sugar-beet farmers, 24 operated a carpet workshop on their premises, of whom about ten used wage labour in addition to family labour (Madjd 1983, p. 27) Moreover, the wages of a young weaver in 1976 were between 400 and 500 rials per day while a skilled weaver would command up to 700 rials per day (Madjd 1983, p. 27) The wages of an unskilled construction worker in the same year were between 350 and 400 rials.[8] Carpet-weaving required little capital since the rug merchants would provide all inputs except labour and would buy the finished product at a fixed price. This meant that, relative to farming, market risks would be fewer.

Persian carpets are unique products (especially as a status symbol) which are not easily substitutible by imports and, although they are exported (traded), they did not face strong import competition. In other words, in the domestic market, the very high income elasticity of demand – plus the fact that Persian carpets are a heavily differentiated product – led to standard Dutch-Disease effects that is, a rise in their price, since they were in effect sheltered from import competition. Moreover, in the export markets the rise in the domestic price of Persian carpets was outweighed by the fact that world demand for these products was rising rapidly. Therefore, strong domestic and international demand maintained an attractive price for Persian carpets in this period. It is clear, then, why carpet-weaving was a competitor for rural labour. Agricultural production in Iran is largely labour-intensive and such a rapid rise in wages, caused by urban demand and rural industries, greatly squeezed that sector. Higher wages meant both higher cost of hired labour and higher opportunity cost of family labour. The higher wage/opportunity cost of labour and institutional difficulties referred to earlier had an adverse effect on agricultural production.

So far the discussion has taken account of the factors which enter investment decisions on the input side. The discussion must now be expanded to cover the price of agricultural products in order to determine the extent to which changes in crop prices affected investment and output in agriculture. In this regard, two types of agricultural product can be distinguished: products which are traded at the margin; and products which are non-traded at the margin. In the first group belong the main cereals, such as wheat, rice, and barley.[9] In the second group belong the more labour-intensive small-farm products such as fresh fruits and vegetables. This distinction neatly fits into the adjustment model of Chapter 1 and it is interesting to note both the price and the output response of these two main categories, in order to assess the strengths and the weaknesses of the economic/policy theoretical framework outlined in Chapter 1. We now consider some of these crops in more detail.

Wheat

Wheat is the main staple food of the population and accounts for over half of the total cropped area.[10] Wheat production was largely stagnant in the 1970s and fluctuation in wheat production is largely attributable to weather conditions. In two good years, 1973 and 1979, wheat production reached about 4.5 million tonnes while on average (excluding these two good years) wheat production was around 3.6 million tonnes (Table A.29). To get an idea of the reliability of these figures, we can use population, per-capita consumption, and import figures. Iran is considered to have one of the highest per-caput consumptions of wheat in the world. Average per-capita use, including seed and other requirements is close to 160 kg (Madjd 1983, p. 30). In 1966–7, with an estimated population of 26 million, total wheat requirements were 4.2 million tonnes, and given 212,000 tonnes of imports, domestic production was near 4 million tonnes. By 1977, population had grown to 36.4 million and, assuming no change in per-capita consumption, total wheat requirements were 5.8 million tonnes, with imports running at 1 million tonnes per year, domestic production appears to have been about 4.8 million tonnes. If we relax the assumption of unchanging per-capita consumption because of income and substitution effects away from wheat, the actual production would be lower. So the maximum figures of 4.5 million tonnes for 1973 and 1979 appear realistic.

Now let us consider the support price of wheat. This is an important price to consider since there was substantial government intervention in the wheat market. A government agency was directly in charge of the importing and marketing of wheat. Also, the government bought a sizeable fraction of the domestic output. According to an FAO (1982a) study on the wheat support price, between 1970 and 1978 the ratio of the volume of wheat procured by the government to total production was 11–13 per cent. Domestic wheat prices were basically adjusted according to the import price. Bread prices in the urban areas were stabilised at a low level by means of a variable subsidy. Urban millers bought subsidised wheat from the government and would, in turn, sell flour to the bakeries at a fixed price.

Looking at the 'support' price of wheat, we find that the government's support price was constant at 6,000 rials per tonne from 1970 to 1973. It then rose steadily, reaching 36,000 rials in 1981(Table A.31). The deflated support price, however, rose only in 1974 and also after 1978, generally falling during the other years of the 1970s. As marketing and production costs soared, reflecting the general rise in the prices of non-traded commodities such as labour, transport and storage, the price fixed by the government became less and less attractive in spite of the stepwise rises in the government's 'support' price after 1974. The trend growth of the wheat price throughout the 1970s was around 14 per cent per year while the growth of unskilled construction workers' wages was about 28 per cent per year, that is, the opportunity cost of rural labour rose twice as fast as the price of wheat (Chart A.5). The gradual erosion of profits in the wheat sector took place via price scissors, that is, lower wheat prices through wheat imports and rising cost of production due to the rising prices of non-traded goods. Given the unfavourable terms of trade facing wheat producers, it is surprising that wheat production remained stable in this period, rather than falling. Moreover, as was mentioned in the last section, the bulk of marketed wheat is produced by commercially orientated medium-sized farms which were increasingly mechanised. Table 4.1 shows the shares of different sizes of land-holding in cereal production. The table does not say that no wheat is produced by the small farmers, but that small farmers mostly produce wheat for their own consumption. It is plausible that the stagnation is partly the result of the declining level of production of small farmers who moved increasingly into fresh fruits and vegetables while larger farmers did not expand the area under wheat due to unfavourable terms of trade.

Table 4.1 Share of cereal production in different-sized farms

Farm Size (Ha)	% under wheat, barley, and rice	% under wheat and barley
less than 1	52	40
1–2	68	49
2–5	82	72
5–10	84	82
10–50	84	83
50–100	62	62
over 100	68	67

Source: Results of the Survey of Agriculture, first stage, 1973, Tehran, 1975, Quoted in S.V. Fallah, 'Mechanisation of Agriculture in Contemporary Iran', Toos pub., Tehran, 1982, p. 126.

One should note that, although the domestic terms of trade between wages and the price of wheat were turned against the latter, the domestic price of wheat was not generally below the CIF import price with the exception of the two years, 1974 and 1975, when there was a surge in the world wheat price (Table A.32). In other words, the existing price protection was insufficient to protect the wheat producer against rising domestic production costs and the loss of comparative advantage.

In addition, harvesting costs had risen substantially due to the loss of labour brought about by the continued migration to the cities, and this affected the dry land areas in particular, which are not mechanised to any great extent. About 50–65 per cent of Iran's total wheat production is harvested by hand (American Cultural Attaché 1977, p. 2).

One reason that wheat production did not experience a greater decline in output may be due to various productivity-enhancing measures taken by both the government and the farmers. Overall fertiliser consumption, using the FAO's (1982b) nutrient content measure, rose from 120,000 tonnes in 1969–70 and 332,000 tonnes in 1974–5, to 389,000 tonnes in 1977–8 and 572,000 tonnes in 1980–1. At the same time, however, there was an increase in the mechanisation of agriculture through the use of tractors, combines, and tillers. The number of tractors in use rose from 30,000 in 1973 to 70,000 in 1979, the number of combines more than doubled, and the number of tillers rose by over 50 per cent. (*Statistical Yearbook of Iran* 1981–2, p. 309).

The conclusion we arrive at concerning wheat is a stagnant production trend and a stable or rising price (except for two years). The evidence concerning price is consistent with the model outlined in Chapter 1, where the terms of trade between traded and non-traded goods turn in favour of the latter. The case of wheat is especially interesting since it is a typical traded commodity, the domestic price of which does not rise with rising domestic demand due to imports. The stagnant price and the rising cost squeezed the sector in spite of the domestic price of wheat being for many years above the world price. Madjd (1983) for example, does not seem to recognise that, although wheat received nominal protection against direct imports, it did not receive protection effectively against rising costs. Also, uncertainties caused by the reversals of government policy *vis-à-vis* small and large farms and the distortions in the land market, created additional disincentives which contributed further to the stagnant production trend. Moreover, government subsidisation of the productivity-enhancing inputs does not appear to have cancelled the adverse impact of falling prices.

Rice

Rice is the main substitute for wheat in the Iranian diet. Domestic rice is considered superior to imported rice, which is cheaper and consumed by most sections of the population. Most of the rice is grown along the Caspian littoral. During the 1950s Iran was an exporter of rice but increased income and population have lead to large rice imports. In the early 1960s per-capita consumption of rice was about 20 kg. By 1977–8 it had almost doubled (Madjd 1983, p. 30). During the 1973–6 period, for example, rice demand increased at a compound rate of 6.35 per cent per year as compared to a corresponding demand increase of 3 per cent per year for wheat in the same period (Centre for Agricultural Marketing Development 1977, p. 18).

There are two rather different estimates for the production of rice in the 1970s in Iran. According to the Central Bank, production was mostly stagnant from 1968 to 1973 with production at little over 1 million tonnes. Production rose by almost 50 per cent in 1973 to 1.5 million tonnes and continued at this level until 1978, declining to 1.1 million tonnes by 1980 (Table A.29). According to the Statistical Centre of Iran, production of paddy increased from 973,000 tonnes in 1973 to 1.13 million tonnes in 1977, from which point it declined (Table A.29). The Central Bank's figures imply a 50 per cent jump in one year, which is rather difficult to believe. Both sources indicate a rise in rice production in this period although the extent of the rise is quite different, depending upon the source. One factor which could have contributed to the rise in production appears to have been the productivity-enhancing policies of the government. A Rice Impact Programme (RIP) was started by the Ministry of Agriculture in 1968 to boost rice production through the supply of improved seed, fertiliser, pesticides and herbicides at subsidised prices. The subsidy on fertiliser and seed was 20 per cent in 1976, while pesticides and herbicides were supplied free of charge (CAMD 1977, p. 6). The fertiliser supplies under RIP increased from about 6,000 tonnes in 1968 to over 50,000 tonnes in 1976. The supply of improved seed increased from 109 tonnes to 416 tonnes from 1968 to 1973. In addition 42 tonnes of herbicides were distributed to farmers by RIP in 1976 (CAMD 1977, p. 6).

Except for imported rice, there was little official price control on local rice in the 1970s. This is in contrast to wheat, the price of which at the farm gate was guaranteed by the government. Annual wholesale price averages of different varieties of rice at Rasht market, which is the major rice distribution centre in Iran, are shown in Table A.33.

Although nominal rice prices rose for most varieties, the price rise lagged far behind both wages and the non-oil GDP deflator. This was in spite of the domestic prices being above the import parity price for the better grades of rice which were equivalent or superior to the imported variety. In fact, champa rice, which is an inferior domestic variety, did not rise in price at all, at least up to 1977. Government intervention in the rice market, by acting as the exclusive importer and distributor (up to 1976) at prices well below the price of local rice, is an important factor influencing the lagging price of rice which, due to the income effect of oil, should have risen substantially. For instance, in the year 1976, the retail price of government-supplied imported rice was 40 rials per kg, compared with 75–80 rials per kg for the most commonly marketed local variety, Sadri Machini (CAMD 1977, p. 23).

This means that the policy of importing to keep down consumer prices did not give sufficient consideration to domestic-producer rice prices and the price development is consistent with rice being a traded commodity, the price of which does not necessarily rise with the rise in domestic demand since rice is being imported. Nevertheless, the productivity-enhancing policies of the government through the RIP apparently had to some extent cancelled the adverse impact of rising costs and allowed for expansion of rice output. Also, most of the rice is produced in the northern provinces of Guilan and Mazandaran which are well endowed with water due to two large rivers, the Sefeed-Rud and Haraaz and several smaller tributaries, thus irrigation has not been an important constraint.

Barley

As for barley, which is mostly used for animal feed, we again observe a stagnant production and a fall in the real producer price due to both rising costs and low prices maintained by large imports (Table A.30, Figure A.5). Again the fall in relative price is consistent with barley being a traded commodity; moreover, the adjustment model would predict a fall in production of such crops, but given the trend towards some technological improvement, a stagnant production may well have been the result. In the case of barley, the domestic support price between 1974 and 1977 was considerably below the CIF import price and it was only after 1978 that the support price was set above the import price.

Barley producers, therefore, were squeezed both by receiving low prices as well as facing high costs. It is surprising that barley output did not experience a sharp decline. One reason may be, though this is personal speculation, that most of the domestically produced barley is not marketed and it is used for on-farm animal feed and the output is hardly responsive to the government purchase price. Interestingly, in 1979, when the new government raised the barley price by 23 per cent, there was no noticeable increase in output (Bank Markazi Iran, *Annual Report and Balance Sheet* 1979, p. 25).

Sugar-beet

Another important and interesting crop is sugar-beet. Sugar-beets are grown in most parts of Iran and it is a labour-intensive crop. Traditionally, the marketing of sugar has been highly regulated. The government would purchase the entire output at a set price and the government is the sole importer and distributor of sugar. Retail prices are also set by the government. The purchase price of sugar showed a trend growth of 16.4 per cent per year between 1971 and 1978 and a trend growth of 15.7 per cent per year if we only take the 1973–8 period (Table A.36). The price change is above the non-oil GDP deflator but not above the rise in wages.

Sugar-beet producers faced a number of advantages in terms of both marketing and credit. The price of the crop was announced prior to the planting season and the factories would supply cash and other inputs at low interest to the producers. Even long-term government capital was made available to sugar-beet farmers through the sugar processors (Nowshirvani 1978, p. 20; Madjd 1983, p. 39).

Sugar-beet is the main source of sugar production. Sugar-beet output gradually increased during the 1970s (Table A.29). It rose from 3.4 million tonnes in 1968, to 4.2 million tonnes in 1973, declined slightly in 1974, rose again to a record of 5.2 million tonnes in 1976, and then declined considerably in 1978 when it fell to only 3.9 million tonnes. The two periods of 1968–73 and after 1973 are therefore quite distinct. In the earlier period output had a rather stable trend growth of 3.6 per cent per year, with R^2 equal to 0.73. The post-1973 period is characterised primarily by strong output fluctuations and if we took the trend growth of the 1973–79 period, we would get only 0.8 per cent growth with R^2 equal to 0.27.

The main reason for the decline in output after 1976 appears to have been the rise in labour and transportation costs. Sugar-beet price and wages stayed in line up to 1974, after which wages began to outpace price rises (Table A.36). Between 1973 and 1977 the opportunity cost of rural labour, that is, the wage rates for unskilled construction work, rose by 273 per cent, whereas sugar-beet prices rose by only 62 per cent.

Moreover, given the quick harvest and perishability of sugar-beet, efficient and timely transport is crucial. During the late 1960s and early 1970s, trucking rates varied between 180 and 400 rials per tonne. Transport costs, a non-traded activity, rose rapidly after 1974. According to Madjd (1983, p. 41), 'during the 1976/77 harvest, competition between the farmers drove transportation rates to as high as 900 rials per ton for a relatively short distance and this was equivalent to one-third of the factory-gate price of sugarbeet'.

Sugar-beet prices remained relatively low because of the high levels of sugar (and honey) imports by the government. Sugar imports rose by about seven times between 1973 and 1975 (Table 4.2). The high levels of sugar imports by the government were the main reason for the relatively slow rise in the price of sugar-beet. Large numbers of sugar-beet producers abandoned the crop despite the marketing and credit advantages of growing it.[11]

Table 4.2 Iran – Sugar (and honey) imports, 1972–9 ($ million)

Year	72	73	74	75	76	77	78	79
Value	26.3	75.2	142.5	542.5	244.9	163.4	246.3	258.0

Source: FAO, *Trade Yearbook* 1978, p. 733; 1980, p. 335.

Clearly the level of protection granted to the crop was insufficient to maintain the existing levels of output. In fact, between 1974 and 1977, world sugar prices rose substantially and the domestic price paid to producers was substantially below the world price in those years (Table A.36). Nevertheless, it is not clear if additional protection would have provided a solution to the problems facing sugar-beet producers. Higher prices would have in any case been quickly overtaken by higher labour costs and, indeed, it would have been difficult to provide the levels of protection required to meet higher costs on either efficiency or equity grounds even in the long term.

Sugar-beet in Iran is not as such traded but it is indirectly so since in its processed form, as sugar, it is traded. Large imports of sugar prevented the rise in the domestic price of both sugar and sugar-beet. Although the domestic price of sugar-beet rose in nominal terms, it declined relative to labour and transport costs. Also, the domestic farm-gate price for sugar-beet was below the world price. The absence of organised channels for the export of sugar-beet, plus legal restrictions on the sale of sugar-beet, meant that the bulk of domestic producers were limited to domestic buyers which were the large public sugar-beet processing plants.

The experience of sugar-beet is a clear example of the political economy of the 'Dutch-Disease'. Government spending policy raises the price of non-traded goods and hence also the cost of production facing producers. The government's import policy prevents the rise in the price of sugar and therefore of sugar-beet in the domestic market. The domestic price paid to the farmers was even below the world price. Even at world prices, however, sugar-beet producers would have had considerable difficulty in coping with the rise in the cost of production. The suddenness of the rise in the cost of production and the maintenance of a relatively low price (to the benefit of consumers) meant that sugar producers were also unable to undertake appropriate technological transformations to lower their unit costs.

Cotton

So far we have been discussing food crops. Now let us consider the main export crop, cotton. Cotton production also appears to have been largely stagnant (Table A.29), except in 1979 and 1980 when it experienced a large decline. Cotton exports, however, which rose throughout the 1960s, declined substantially in the 1970s. In other words, production was largely stagnant while exports declined especially after 1975 (Table A.30). It follows that the domestic market was absorbing an increasing proportion of the domestic production. Furthermore, international cotton prices in the 1970s were not rising, which means that the domestic cost of production must have rapidly eliminated the profitability and competitiveness of Iranian cotton. Therefore, the cotton producers turned towards the domestic market for the sale of their output since they would at least have an advantage in terms of transport costs.

Other non-cereal crops

Looking at some of the other non-cereal crops, some interesting trends emerge. Table A.38 shows the output of seven important non-cereal crops in the 1970s. The rapid growth in production of onions, potatoes and tomatoes contrasts with the slower growth in production of oilseeds, tobaccos, pulses and tea.

Tea. The tea crop between 1972 and 1976 was stagnant. The figures for 1979 and 1980 refer to dried leaves and are not directly comparable. Using a 50 per cent conversion ratio for fresh to dried leaves in terms of weight would bring the tea crop to 68,000 and 86,000 tonnes for 1979 and 1980, respectively. This is still a stagnant level as compared to the early 1970s. Tea imports in this period rose considerably, from 6,200 tonnes in 1970 to 22,000 tonnes in 1980, with a sharp jump from 7,500 to 12,500 tonnes in 1974 (Table A.38). This shows that tea was largely traded in this period; tea imports prevented a rise in the domestic price sufficient to expand domestic production, resulting in stagnation.

Oilseeds. Comparing the 1970–4 and 1974–9 periods, a sudden rise in the average level of oilseed output occurred. The average for the 1970–4 period is 58,800 tonnes while the average for the 1975–9 period was 112,000 tons, twice as much as in the previous period. Taking the 1970–80 period, the cumulative growth of output of oilseeds was 20 per cent for the period as a whole. Oilseeds had a ready market in the domestic oil processing industries which were shown in Chapter 3 to have experienced rapid growth in the 1970s, as a result of the high domestic demand and a highly mechanised prod-uction process due to the large IMDBI loans (see Tables A.21 and A.23). In other words, the growth in oilseed production was derived from the externality generated from the growth in vegetable-oil processing. Also, the volume of vegetable-oil imports was small because of the presence of a large domestic capacity. The case of oilseeds is therefore, an example of a non-traded crop (inside the cone in Figure 1.1 of Chapter 1), which had considerable comparative advantage even at changed relative prices, largely due to the presence of a strong domestic processing industry. It was clearly cheaper to buy from the domestic producers than to import the oilseeds necessary for crushing, or buy the ready vegetable oil. Moreover, although it has not been possible to find a documented study of oilseed production in Iran, the author himself is familiar with some of the major areas of oilseeds production in Iran which are largely located in the Maz-andaran region (north-east). As was mentioned earlier, the Mazandaran region is agriculturally one of the most commercially developed and mechanised areas. Large mechanised commercial farms are the main producers of oilseeds. This is another reason for the growth in oilseeds in the 1970s, in spite of the highly tradable nature of the crop both in unprocessed and processed form.

Tabacco. The cumulative growth of tobacco production was zero if one takes the 1972–80 period, although changing the base year to 1971 would give a 26 per cent cumulative growth for the period. In most years during the 1970s, however, tobacco production was below the 1971–80 average of 18,000 tonnes. In other words, tobacco was a lagging crop. The tobacco crop is bought mostly by the tobacco monopoly, which is a government enterprise. In this period, little raw tobacco was imported for processing by the tobacco monopoly. Imports of manufactured tobacco, however, then rose rapidly because of purchases by the monopoly itself. Table 4.3 shows the production of domestically manufactured and imported cigarettes in the 1970-7 period. As can be seen, imports of cigarettes by the monopoly itself rose from zero in 1973 to 10.6 billion cigarettes in 1977, which is equivalent to nearly 75 per cent of domestic production. Domestic production is stagnant during the 1973–7 period, which is a period of rapidly rising demand. The stagnant level of domestically manufactured tobacco resulted in a stagnant level of demand for domestic production of tobacco leaves. In other words, tobacco becomes indirectly traded through the imports of processed tobacco, hence the stagnation of tobacco production. Tobacco, therefore, is another good example of the adverse impact of the adjustment policy (through higher imports) on a traded, agricultural crop.

Table 4.3 Iran – Production of domestically manufactured and imported cigarettes   (billions)

	Domestically manufactured	Official imports
1970	11.9	0
1971	13.4	0
1972	12.8	0
1973	14.8	0
1974	14.4	1.1
1975	15.2	3.0
1976	15.2	6.8
1977	14.3	10.6

Source: IMDBI (1970–8)

Pulses

The cumulative growth in the production of pulses was 11 per cent between 1970 and 1980. During this time, in five years out of eleven, gross output of pulses was above average and their overall performance was sluggish. Imports on average accounted for less than 5 per cent of domestic production and competition from imports did not appear to be a significant factor in the sluggish performance; though in 1977 and 1978, when there was a sudden rise in imports, production appears to have suffered. The supply side, therefore, does not seem to provide an adequate explanation for the sluggish output. What about the demand side? Pulses are generally agreed to have a low income elasticity of demand. This largely appears to account for the sluggish domestic output. In fact the combined volume of domestic production and imports was fairly stagnant, reflecting the absence of a large incremental demand as expected from the rise in the incremental income after 1973. In four years out of eight between 1973 and 1980, total quantity demanded (produced and imported) actually declined, while in other years it generally grew slowly relative to income (Gahreman 1983, p. 215).

Potatoes, onions and tomatoes. Now let us consider the fast-growing crops, namely potatoes, onions and tomatoes (Table A.38). These crops had very rapid rates of growth with above average output in the latter part of the 1970s. Imports of these products, with the exception of onions in 1974 and 1976, have been negligible. The rise in domestic demand resulted in high prices for these crops. Potatoes, onions and tomatoes are mostly irrigated field crops produced by the small farmers using family labour, with its lower cost. The rising demand in the cities provided a ready market for them. The bulkiness of these crops and the problem of perishability in the case of tomatoes, prevented imports of these crops. The author's own field observations during the summers of 1975 and 1977 also confirmed the rapid rise in the urban periphery production of carrots and fresh greens. The evidence, although somewhat limited, gives support to the generalisation that there was a shift within crop production away from cereals and cash crops and towards the production of *non-traded* root crops and vegetables; also bulky fruits, such as melons, are not traded, while there are considerable imports of oranges in this period.

Conclusions

The conclusion that emerges so far is that the relative price model presented in Chapter 1 performs fairly well in explaining the relative price movements in the 1970s. Moreover, it has been shown how there was a general stagnation of internationally traded crops in this period while non-traded crops such as vegetables experienced considerable growth. Moreover, traded crops with a mechanised production process, such as oilseeds, also did well. Commodities such as cotton, which was an export crop, declined when Iranian cotton became non-competitive for export purposes; the country began to import not cotton but cotton products such as textiles. In other words, the country's comparative advantage changed very rapidly with the changes in the ratio of output price to costs in the various sectors of the economy. Moreover, although food crops were generally priced above the CIF import price, the rising costs of labour and transport resulted at best in their stagnation. The rapidity of the change in the price/cost ratio plus the predominance of small farms, in such cases as sugar and rice, inhibited mechanisation which could have offset the adverse movement in labour costs.

Production performance, however, is not adequately explained by the model since supply responds to both price and non-price measures, and although the falling real price may explain one part of the stagnant agricultural production, it is not by itself a sufficient explanation. Other factors, such as the government's pricing policy, investor uncertainty concerning property rights, and technological change, must be explicitly introduced into the model in order to arrive at a more satisfactory explanation of production performance. For example, the movements of relative prices in agriculture were very much influenced by the government's agricultural price policies. The existence of a cheap-food policy in Iran, maintained by large food imports, meant that food prices did not rise as much as they would have, given the higher demand for food as the result of the spending effect of the oil income. It may well be argued, however, that higher producer prices would not have made that much difference given the time-period in which wages and marketing costs soared. Indeed, the short-term solution, above all, was to adopt a much slower rate of absorption of oil revenues. This could increase the demand for non-traded goods in line with the increase in the supply of such goods. Also, the government could have facilitated greater access of the agricultural sector, and especially of small farmers, to credit, encouraging productivity growth. Indeed, the long-term solution to the problems of Iranian agriculture would have required productivity-enhancing policies to reduce input costs as well as a number of institutional adjustments such as a better definition of property rights.

The key conclusion from studying the agricultural sector is that the relative price demand-based adjustment model of Chapter 1 describes the experience of crop prices in the agricultural sector fairly accurately, although one should not overlook the role of the government in actually permitting or facilitating international trade in these products to take place. The stagnant price of traded crops and the rising cost of production resulted in stagnation or decline of such crops while the rising prices of non-traded agricultural crops resulted in the growth of output in that sub-sector. Putting the traded and non-traded crops and their price levels in Figure 1.1 of Chapter 1, we can show how various crops move in a spectrum from being exported, through non-traded to imported, if the pre-boom and post-boom periods are compared. A transition which reflects, among other things, the interaction between the real appreciation of the exchange rate and government policies towards the agricultural sector (Figure 4.1).

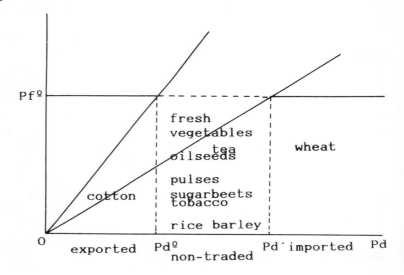

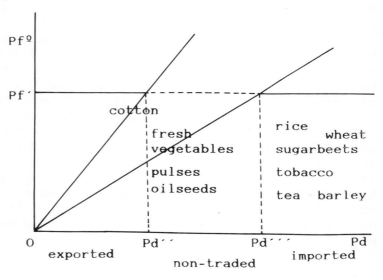

Figure 4.1

Figure 4.1 clearly shows how a fall in the relative price of traded goods from P_f'' to P_f' changes the areas of the squares which define the zones in which crops are exported, imported, or non-traded. After 1973, cotton remains an export crop although exports gradually fell to zero by late 1970s and most of the production was being used at home. A number of commodities such as oilseeds, tomatoes, fresh vegetables, and pulses remained non-traded. Finally, tobacco, tea, sugar-beet, and rice entered the imported zone due, among other things, to relative cost advantages abroad and higher incomes at home. Figure 4.1 therefore sharply demonstrates the adjustment process as it affected crop production and trade in the agricultural sector.

What is, however, more interesting and somewhat puzzling is how the experience of traded goods in the agricultural sector is different from the experience of industrial sector. As has already been discussed in Chapter 3, there were a number of fast-growing industries producing traded goods in the post-1973 period. These were industries which were mostly highly dependent on imports for their capital inputs, which also meant that they had higher capital intensity than other industries. Contrary to the experience of these fast-growing industries which were producing traded goods, almost all traded agricultural products experienced stagnation or decline. One answer to this puzzling development which emerges from the analysis and examples so far lies in the relative creditworthiness of these two sectors and their differential access to credit and technology.

The Industrial and Mining Development Bank of Iran plus a host of measures to boost the industrial sector (such as tariff protection, subsidised inputs, cheap credit) which began in the 1960s, meant that a number of industries with continued access to official support, could withstand the 'Dutch-Disease' much better than the rest of the traded sector. More specifically, the agricultural sector inherited a fragmented and weak production structure due to the land reform and the resultant uncertainties in land rights and government policy regarding the sector. This, and the pro-industrial bias of the government, resulted in the increased vulnerability of the agricultural sector to rising costs, especially in traded crops such as wheat. The agricultural sector, therefore, suffered from a lack of creditworthiness, hence lower capital intensity and less ability to respond to the 'Dutch-Disease'. The examples of modernisation in the agricultural sector were few, negligible, and generally ineffective.

The relevance of the institutional analysis of this chapter should now become evident. The land reform, by increasing the fragmentation of the land and creating a mass of poor peasants, resulted in lower creditworthiness and lower capital intensification in the agricultural sector.

To summarise, this chapter has not only attempted to explore the process of adjustment to relative prices, but has also shown, from the available data, how supply response is more complicated than being merely a function of relative prices. Both the government's policy and the existing agrarian structure, that is, institutional factors, have been brought in as major factors underlying the supply response. This result has important implications not only for the 'Dutch-Disease' model, but also for the economic adjustment models as a whole.

A typical example of how many analyses of adjustment do not explicitly consider such factors is the approach of the International Monetary Fund to structural adjustment in the developing countries. Devaluation of the exchange-rate, which is the major policy advice of the IMF for countries in balance-of-payments difficulties, assuming it sticks, is only a temporary relative-price manipulation. That the actual supply response of traded goods is a far more complicated process and that the production of traded goods must be put within the overall capital utilisation in the economy is hardly ever emphasised. In fact, critics of the IMF such as Killick (1983) have referred to the absence of supply-side analysis in the IMF's official thinking.

Notes

1. In some parts of the country such farming groups have a different name such as sahra, taaq, or haraaseh. For a detailed analysis of the *boneh*, see Safi-nezhad (1977); for a discussion of the *boneh* in English, see Hooglund (1982, pp. 22–8).
2. Ministry of Rural Affairs and Co-operatives, *The Eleventh Anniversary of the Passage of the Land Reform Law*, Tehran, 1972, p. 33, cited in Madjd (1983).
3. Hasan Arsanjani, *The Land Reform,* Tehran, 1959, p. 198 cited in Mo'meny (1980, p. 380).
4. This figure is likely to be too low, but in the absence of any rural labour survey it is difficult to dispute it.
5. A third measure, the introduction of production co-operatives, was also initiated but these were few in number and, due to the lack of data on their performance, the discussion here is limited to the rural co-operatives and farm corporations.
6. W. Ladejinsky, 'Note on Iran's Agricultural Corporations and Agro-Industries' paper submitted to the Plan Organisation of Iran, Tehran, 1970 cited in Ashraf and Banuazizi (1980 p. 40).
7. Figure A.3 shows the growth of construction related industries, which include brick-making.
8. Calculated from *Statistical Yearbook of Iran* 1980, p. 702.
9. Difficulties, or the insufficiency of imports, could also turn these staple cereals into non-traded ones *at the margin*.
10. Estimated from total cultivated area and area under wheat for 1973 from FAO (1974).
11. M.G. Madjd, *Instability of Sugar Production in Iran,* Plan and Budget Organisation, Tehran, 1981 (in Persian), cited in Madjd (1983).

PART III
THE CASE OF NIGERIA

Introduction to part III

In this part the analysis follows broadly the main outline of Part II. In Chapter 5 the macro aggregates are discussed. Here, the processes of the rising level of expenditure are traced through the analysis of the rising level of domestic deficit expenditure by the government and the components of this expenditure. These components include monetary growth, credit expansion, a boost in aggregate demand and the rising prices of non-tradeables and wages, differential growth in the GDP, and the supply response of output to price.

Chapter 6 is a more detailed analysis of agriculture. In this part, the manufacturing sector has not been treated in a separate chapter due to the paucity of data on the subject, as well as the author's more limited familiarity with Nigeria. Chapter 6 concentrates on the agricultural sector performance both in food crops and in export crops. Moreover, both institutional and policy factors are brought in as major forces affecting the response of agriculture to the adjustment in relative prices and higher levels of income.

It should be noted that Nigeria is larger than Iran in terms of population but it has a lower per-capita income, a smaller manufacturing base, and lower life expectancy. In 1981 the population of Nigeria was 87.6 million, while that of Iran was 40.1 milion (World Bank 1983). In the same year per-capita income of Nigeria was $870 while in Iran it was about $1750. The share of manufacturing in the total GDP of Nigeria was 6 per cent, while in Iran it was 9.5 per cent (see also Tables A.8, A.9). In 1980, life expectancy in Nigeria was 49 years old, while in Iran it was 58 years. If one considers per-capita income, the share of manufacturing in GDP and life expectancy at birth as indicators of development, it can be said that Iran is a more developed country than Nigeria.

Moreover, in 1973 Nigeria had just begun to export oil while substantial oil production in Iran goes back for several decades. In Nigeria, therefore, it was increased production combined with higher prices which affected the economy while in Iran, as we saw, oil production did not change significantly in the 1970s. This means that the impact of the oil boom on the Nigerian economy involved a combination of higher price and a greater level of production. As in the case of Iran, oil production would be regarded as an enclave sector for Nigeria and oil revenues as transfers. The larger relative spending impact combined with a less developed economic base, meant that the oil boom impact on Nigeria is likely to have been more severe in terms of pressure on domestic supply as a result of higher demand generated by the oil income.

Furthermore, Nigeria was before the oil shock an economy based on non-oil agricultural exports. Iran, on the other hand, did not have substantial non-oil exports before 1973. The impact of the oil exports on non-oil exports in Nigeria was, as we shall see shortly, more significant than in Iran which, as we saw, did not have significant non-oil exports if we leave aside the comparatively minor exports of cotton.

5 Macroeconomic adjustment

Oil revenues and changes in aggregate expenditure

Let us begin by looking at the rise in the foreign exchange earnings of the economy and the consequent rise in government revenues. As the oil price rose, the foreign exchange earnings of the economy rose by over 600 per cent in nominal terms, from $3.6 billion in 1973 to almost $25.3 billion in 1980 (IMF 1983, p. 393). The share of oil in total export earnings rose from 58 per cent in 1970 to almost 97 per cent in 1980 (Table B.1). For Nigeria, as Iran, the current account and oil sales are the key to the balance of payments; the capital account has basically an accommodating function, as in borrowing or lending abroad in order to facilitate the balance in the current account. Moreover, speculative capital inflows are absent since the capital market is insufficiently developed and there are many restrictions on capital movements.

In Nigeria most, but not all, oil earnings accrue to the government, although the government's share steadily increased throughout the 1970s. The tax on each barrel of equity crude (the government share) rose from $0.90 out of $2.25 (40 per cent of total sales) in 1970, to around $30 out of $35 in 1980 (85 per cent of total sales) (Kirk-Greene and Rimmer 1981, Ch. 7).

In other words, government revenues rose more than proportionally to the oil price rise since government's share in total oil sales also increased in the same period. This gave the government a key role in determining the form of adjustment. Total government revenues rose from 1,695 to 10,912 million naira or, a compound growth of 30.5 per cent per year between 1973 and 1979 or 10.5 per cent per year in real terms (see Table B.14 for the deflator). The share in total federal revenue of petroleum profit tax and mining rents and royalties (mostly from oil mining) rose from 60 per cent in 1973 to 80 per cent in 1979 (Table B.2). In the same period non-oil revenues rose at about 16 per cent per year or about one-half of the growth of revenues due to oil. Export duties in particular declined spectacularly, from 41 million naira in 1970 to 0.2 million naira in 1979.

In Nigeria, much as in Iran, the 'non-adjustment' option has not been significant, while among the three adjustment options discussed in Chapter 1 — the explicit appreciation of the exchange rate, greater domestic absorption, and lower commercial barriers — the second appears to have been the most important.[1]

Considering the nominal variations of the exchange rate in the 1970s, if anything, the naira has actually been depreciating in nominal terms. Only in 1976 and 1977 did the naira appreciate slightly (less than 1 per cent); otherwise, between 1970 and 1980 the currency depreciated nominally against the US dollar by over 24 per cent.[2] In other words, the nominal appreciation of the currency has not been used to any extent for adjusting to the accumulation of reserves in Nigeria.

Now consider the option of adjustment through higher absorption — greater expenditure on both domestic and foreign goods and services. Looking at the growth of imports of Nigeria in the 1970s, it becomes clear that the greater foreign exchange earnings of the post-1973 period have been primarily used to finance a much higher level of imports into the country. Indeed, imports have tended to outpace exports in most years (Table B.3). Between 1973 and 1980, exports rose at 28 per cent per annum, while merchandise imports grew at 32 per cent per annum. Overall current account payments, which also include payments for services, grew at 24.7 per cent per annum. Overall current account payments were generally greater than total merchandise exports throughout the 1970s. The point that clearly emerges from these figures is that Nigeria has chosen to adjust by greater imports while maintaining a relatively stable nominal exchange rate.

Furthermore, some of the earlier restrictions on the import of consumer goods were slightly relaxed. Table B.4 shows the changes in nominal protection afforded to the different branches of the Nigerian industrial sector between 1957 and 1979. Comparing 1967, which was a pre-boom situation, to 1977, three years after the boom, the relaxation of tariffs after the boom on many of the industrial commodities is readily observed. From the 26 commodity groups for which data are available, 22 experienced a lowering of tariffs of between 12 and 100 per cent. The remaining four, whose tariffs remained constant or increased, are wood products, industrial chemicals, iron and steel products, and certain other miscellaneous metal products. Interestingly, comparing the 1979 level with the 1977 level or protection, there was a general increase in tariffs for over 50 per cent of the commodity groups for which data are available. The first half of 1979 was a post-boom period when the balance of payments surplus had turned to a deficit with a general tightening of expenditure and imports, a point which will be discussed later in the chapter. The main emphasis at this stage however, is on the adjustment to the boom and it is the lowering of the tariff during the boom which is of immediate interest. This is because it shows that partial lowering of the tariff barrier was among the instruments used for adjustment to the balance of payments surplus. Nevertheless, in Nigeria as in Iran, the main instrument of adjustment after the 1973 rise in foreign exchange earnings is higher levels of domestic expenditure, fuelled by rising levels of government expenditure.

The government's domestic deficit had suddenly risen in 1974–5 and 1979–80 in response to the two oil price rises, while between 1975 and 1978 it was rather stable. Taking the 1973–80 period as a whole, however, the domestic budget deficit grew at 41 per cent per year in nominal terms and 25 per cent in real terms (using the CPI as a deflator) (Table B.5). The size of the domestic budget deficit was about 13 per cent of the total GDP, averaging around 12 per cent of the GDP and 20 per cent of the non-oil GDP. The deficit grew from being a quarter of total exports to around a half from 1973 to 1980. The size of the deficit is therefore considerable, as was also the case for Iran. Along with the domestic deficit, the money base and the money supply also grew rapidly. Figure B.1 shows the domestic deficit, the supply of money (M2), and the claims of the banking system against the private sector. Of course, the claims against the private sector have to be counterbalanced with the private sector's balance-of-payments deficit before we can arrive at the true liquidity impact of the private sector. Unfortunately, the Nigerian government does not publish separate figures for the private and the public sector's imports and exports. Moreover, due to this lack of data, the calculation of the domestic deficit has been through an indirect method.[3]

Using the derived deficit figures and credit to the private sector, we can calculate the impact of the domestic deficit and the credit to the private sector on the money supply. Given the limited number of observations – the eight years 1973–80 – and the strong possibility of autocorrelation between these variables, we have used the first differences of these variables and regressed the money supply on the domestic deficit and credit to the private sector. This gives:

M2 = 18.9 + 0.273(DBD) + 0.119 (CPS)
(0.82) (2.32) (0.19)

where M2 is the percentage in the supply of broad money,
DBD is the percentage change in domestic budget deficit and,
CPS is the percentage in change credit to the private sector;

$R^2 = 0.60$, $R^2 = 0.40$ (adjusted for degrees of freedom), and the Durbin–Watson statistic is 2.26. Only the percentage change in the budget deficit is significant at the 99% level of significance as a determinant of the money supply while credit to the private sector is insignificant, understandably so since we have not accounted for the private sector's balance of payments. Running the regression only with the money supply and the domestic deficit, we get almost the same coefficient for DBD' a similar level of significance and a similar value for R^2.

$M2 = 22.8 + 0.279(DBD)$
$(2.4) (2.7)$
$R^2 = 0.59$,
$R^2 = 0.51$ (adjusted for degrees of freedom), and the Durbin–Watson statistic is 2.19.

The domestic budget deficit therefore accounts for about half of the changes in the money supply and, given its large impact (larger than the impact of the deficit in Iran), it probably has a greater influence on the money supply than the private sector's net liquidity contribution.

Having discussed some of the main determinants of the money supply, we can now discuss the relation between the growth of the money supply and the growth of aggregate expenditure. Here again we need to look at how the money grew faster than money demand.[4]

Money supply grew at 32 per cent per year between 1973 and 1980 (Table B.5). Real, non-oil GDP grew at 3.9 per cent per year (Table B.5), giving an estimated money demand growth of 3.6, 4.6, and 4.9 per cent per year depending upon which elasticity we use. Consequently, assuming that real money demand was a stable function of real income, money supply must have been growing almost 27 or 28 per cent faster than the real money demand. The non-oil GDP deflator however, was growing at 11.9 per cent (Table B.8) and consumer price index was growing at 16 per cent per year (Table B.14). In other words, assuming that the data are reliable, the constant velocity assumption which assumes a stable money demand function does not give us any significant results. Nevertheless, much as the case of Iran, the Nigerian evidence is not inconsistent with the money supply growing faster than money demand (which may or may not have been a stable function of output growth in the non-oil economy), and this results in the rapid growth of expenditure. It should be added that at the initial stages of the boom it is the supply of money which grows as a result of the rising level of public expenditure. The disequilibrium level of many holdings in relation to expected income, therefore, induces expenditure by the receiver of money.

Private consumption expenditure rose at 6.6 per cent per year, government consumption expenditures at 11.3 per cent, and investment or capital formation at 15.8 per cent, all in real terms (Table B.6).

The growth in expenditure in both Iran and Nigeria was indeed among the highest in the world (World Bank 1983, p.154). In both cases the growth in expenditure was only partly financed by oil revenues in so far as the expenditure could be satisfied through imports. The oil-financed expenditure is inflationary from the point of view of non-traded goods. Also similar to Iran, the domestic inflation cannot be accounted for by imported inflation, for example, the US CPI, which only rose at 6 per cent per year (Table B.14). In contrast to Iran, Nigeria has not followed a cheap-food policy and, therefore, the rising price of food has been an important factor in the Nigerian inflation. Table B.14 also shows the relative food price and CPI changes. From 1973 to 1981, food prices rose 32 per cent faster than the CPI.

The rapid rise in expenditure and of the price level without the corresponding changes in the exchange rate resulted in the real appreciation of the naira. Measuring the real appreciation of the exchange rate through the method of correcting the exchange rate for the differential rates of inflation between countries, we find that the naira appreciated against the US dollar by 113 per cent between 1973 and 1980 (Table B.14).

Having discussed the monetary process of adjustment and the rising level of aggregate expenditure at the macro level, it is also necessary to discuss the pattern of government expenditure which fuelled the process. As was argued in the initial chapter it is not only how much the government spends but also how the government spends its higher revenues which affects the process and the outcome of adjustment. At this point, therefore, we examine the pattern of government expenditure which is at the basis of the boom, a pattern which is determined by the government's development strategy as well as by crisis responses stirred by earlier developments.

The pattern of government expenditure

From a base of 100 in 1972–3, the index of Nigerian federal government spending rose to 1,205 in 1977–8 (Table B.7). The rapid rise in the expenditure reflected a great haste in spending the revenues. Interestingly, the first impact of the sudden rise in revenues in 1974 was a vast increase in the government's financial assets with the banking system since it took the whole of 1974 before the spending spree started. Government deposits with the Central Bank rose from 19 million naira in 1973 to over 2 billion naira in 1974. In other words, throughout 1974 over 90 per cent of the funds were held in highly liquid form (Central Bank of Nigeria 1983, p.19). Spending essentially began with a massive wage and salary increase to the civil service, the famous Udoji wage awards. Civil service wage increases ranging from almost 100 per cent at the top of the scale to 130 per cent near the bottom became effective as of 1 January 1975. This was followed quickly by generally comparable increases for the armed forces, public corporations, other parts of the public sector, and also in the modern private sector. The civil service increases were backdated to 1 April 1974 and the backpay was paid as a lump sum (Schatz 1977, p. 31). Table B.7 shows the federal government's index of current expenditure which escalated from 100 in 1972–3 to 443 in 1975–6. Also, the share of current expenditure rose from 33 per cent of total federal expenditure to an average of 60 per cent during 1974 and 1975.

While wage increases were a quick way of spending some of the money, other major projects were rapidly adopted and approved. The Third Plan for 1975–80 quickly became a formality, and new allocations and projects were quickly added with little investigation or appraisal, in a similar way to the confusion which was described regarding planning in Iran. Disregard for the limits to the efficient expansion of domestic investment, neglect of issues relating to project co-ordination and interrelation, and unrealistic increases in the projected levels of output became characteristic of the planning process in this period, while few safeguards were established against waste and corruption (Schatz 1977, p. 49). The index of federal capital expenditures escalated from 100 in 1972–3 to 335 in 1975–6 (Table B.7). The government adopted new measures to ease or circumvent executive-capacity problems in order to raise the rate of investment. Competitive bidding on construction and other contracts was to be dropped or moderated particularly for implementing high-priority projects over a limited period of time (National Planning Office (NPO) 1976, p. 399; Schatz 1977, p. 49). Some contract awards were to dispense with prior appraisal altogether and the design and construction were to proceed simultaneously. Reviews of proposed projects were to be quickened; if a ministry[6] fails, within four weeks, to react to a memorandum on capital projects, it will be presumed that the ministry has no comment and the project would be awarded' (NPO 1976, p. 397, Schatz 1977, p.50 (1977, 50)). Schatz adds:

> Other methods of increasing executive capacity were also adopted. In a move to raise internal incentives and pressures, public agencies and their officials were put on notice that they would be judged primarily by their plan-implementation ratios. Cumbersome procedures for land acquisition were simplified. The assistance of foreign governments was to be invoked in negotiating contracts with foreign construction companies. Restrictions on the use of foreign personnel were to be moderated. Reliance on foreign investors was to be increased, particularly in partnerships with Nigerian government in large industrial ventures.

Meanwhile, substantial bottlenecks were building up due to a limited port capacity, limited transport facilities, and limited manpower. The famous cement scandal in 1975 is a case in point (see *Sunday Times*, 12 October 1975).

By 1976 the balance of payments and the government budget were already in deficit and by 1979 the federal government's expenditure index was heading towards the 1,000 mark. Also there was a considerable underestimation of the costs of projects and the government's surplus position was reversed by the end of 1975. The irreversibility of many of the capital investment projects and the committed contracts resulted in an overshooting of current expenditure. As the budgetary and balance-of-payments deficits grew alarmingly, the government had to resort to restrictive measures. Tighter monetary policies, import controls, increased taxes and various direct measures to control prices were instituted. In March 1979, General Obasanjo in his 1979–80 budget speech announced that the era of easy money was over (*West Africa*, 9 April 1979), and a year later President Shagari resorted to drastic measures to 'restore sanity' to the economy (*West Africa*, 31 March 1980; Schatz 1981, p. 37). The earlier relaxations of tariffs were also tightened and, as Table B.4 shows, there was a general increase in the level of nominal protections. The further rise in the price of oil in 1979, however, came to the rescue of the Shagari government. In his world press conference after 100 days in office, on 8 January, 1980, Shagari declared:

> We have, during the last 100 days, witnessed a tremendous improvement in our economic situation. When I assumed office on October 1st 1979, the daily cash balance of the federal government with the Central Bank was in the red to the tune of 521,748,000 naira. On that date the country's reserve stood at 2,376,824,653 naira. Within three months of our coming into office we have been able to reverse this unhealthy financial trend which had plagued the federal government since July 1976.

Of course, what President Shagari did not mention was that it was not he that had reversed the financial position of the government but the rise in the price of oil from $19 to $34 between April 1979 and February 1980. Few lessons had been learned, however, and a new spending spree started both in the form of rising salaries and through new capital projects, among which the most conspicuous and expensive are the new federal capital, Abuja, the housing schemes, the large irrigation projects and steel complexes. Massive new expenditures and the decline of oil prices after 1981 meant that by the end of 1983 the country's total foreign reserves had dwindled to a little over 900 million naira; this was sufficient to cover only one month of imports if the 1982 level of imports is used as a reference.[5] Again, soaring prices and rising deficits meant new cutbacks and austerity measures. Rising discontent over economic management and rapidly accumulating foreign debts proved to be too destabilising and, by 1984, Shagari's regime was overthrown by a military coup. It appears that, apart from the need to diversify away from oil, due to its non-renewability and volatility as a source of revenue, a substantial cost in terms of political instability has accompanied the hurried spending of oil revenues in both Iran and Nigeria; this is a further argument which reinforces the need to adopt a very different pattern of expenditure, one which generates greater economic stability could in principle contribute to better political stability.

Now consider the composition of capital spending by the government from Tables 5.1 and 5.2 show the functional distribution of public capital expenditures in the 1970–4, 1975–9 and 1980 periods for both the federal and state governments.

Table 5.1 Percentage distribution of actual public capital expenditure (1970–4)

Function	Total, all government	Federal government	State governments
Agriculture	9.0	5.0	14.8
Manufacturing and mining	5.8	3.1	6.8
Education	11.0	7.7	15.6
Health	5.0	3.1	7.2
Transport and communication	25.6	29.0	20.9
Power	5.0	8.0	0.9
Other services	17.3	13.7	23.0
Administration, defence, and security	21.3	30.4	10.8

Source: NPO (1976, p. 26).

The functional distributions of public investments have remained largely constant in the post-1975 period, indicating the broad continuity of public policies in this period in spite of the several changes in the government. The first point to note between the pre-oil and post-oil pattern of expenditure is that the major increase in the share of investment has gone to transport, communications, power and a small increase in the share of education, while agriculture, manufacturing and mines, and health have had a reduced share. In both periods the share of infrastructure and services is considerable, and in the post-oil period about 80 per cent of all public investments are in physical and social infrastructure and public services, the principal components of which are transport (mostly roads), communication (mostly telephones), education and administration (including defence and security). Infrastructural investments are investments in the non-traded sector, and this confirms the observation which was made in Chapter 1 that public investment in that direction substantially reinforces the movement of resources towards non-traded goods. This flow is in addition to the investment flow towards non-traded goods by the

Table 5.2 Percentage distribution of actual public capital expenditure

Function	Total 1975–9	Federal 1975–9	State 1975–9	Federal 1980
Agriculture	7.1	2.4	10.5	5.1
(irrigation)	2.8	3.5	0.7	5.4
Manufacturing and mining	13.7	16.7	4.3	15.0
Education	10.2	7.7	17.8	9.0
Health	2.0	1.2	4.7	2.3
Transport and communication	29.3	33.8	14.7	28.0
Power	5.9	6.9	2.4	6.0
Other services	7.6	4.0	30.9	9.3
(Housing)	4.1	3.8	4.9	4.1
Administration defence, and security	17.3	20.0	9.1	15.8

Sources: 1975–79 from NPO (1981), 1980 from *Economics and Social Statistics Bulletin*, Federal Office of Statistics, Lagos, 1984, p. 31.

private sector, a point which is discussed in the next section. In the 1981 and 1982 approved capital expenditure, two new entries were added. These are Abuja, the new federal capital, and steel was separated out from the rest of the manufacturing sector. Abuja was allocated 2.5 per cent of the approved capital expenditures for 1981 and 1982 and steel was given and approved allocation of 10.6 per cent (Awojobi 1982). The allocation of 2.5 per cent of the government's budget for the establishment of a whole new city is a supreme example of how public funds are diverted towards the production of non-traded goods.

Of the 20 per cent allocated to agriculture and industry, the bulk is in large projects. In industry, almost half of the allocations are for the build-up of the steel industry while small scale industries have a planned allocation of less than 5 per cent in the Fourth Plan (NPO 1981, p.148). In spite of the lack of encouragement by the government, the number of small-scale industries grew at an average rate of 32 per cent per year between 1974 and 1978 (NPO 1981, p.146). This is an important point concerning the type of development which is encouraged by government policy. Expenditure is largely concentrated in the non-traded and large expensive projects in the traded sector of the economy while the small-scale sector is badly neglected; it is the small industries which are the most domestic-resource-based of all industries and have a great potential for viable industrial diversification.

Concerning the agricultural sector, a few broad observations are also in order. Of all the public investments in 1975–79, agriculture received only 7.1 per cent of all allocations. If irrigation is also included under agricultural investment, then the share goes up to 9.9 per cent. Looking at the federal government's allocations alone, the share drops to 2.4 per cent (5.9 per cent with irrigation) while state governments give greater priority to this sector, allocating a little over 10 per cent of their budgets. In 1980, although the share of agriculture in the federal government's allocations doubled, reaching 5.1 per cent (10.9 per cent with irrigation), it still received a relatively small amount, reflecting the fact that agriculture is not a priority sector. Again within agriculture, the bulk of investments was in large-scale food crop projects and large-scale irrigation. Interestingly, the share of irrigation alone was larger than the rest of agriculture combined in the 1980 actual capital spending of the federal government. Moreover, the government's agricultural spending programme involved an increased direct government entrepreneurial role in agriculture which corresponded with a disproportionate allocation of funds and know-how to public and agricultural projects. Also, over 80 per cent of so-called agricultural investment by the federal government was in fertiliser procurement and distribution, reflecting a highly skewed pattern of expenditure, a point which will be further discussed in Chapter 6. The policy and expenditure background broadly described, let us now consider the relative price and output developments for the traded and non-traded goods as reflected in Nigeria's national accounts.

The Gross Domestic Product and its components

Between 1973 and 1981, GDP in current prices rose (using the best fit of a time trend) at 15.6 per cent ($R^2 = 0.90$) per year and the non-oil GDP rose at almost the same rate of 15.7 per cent ($R^2 = 0.92$) per year, which is somewhat puzzling. Unfortunately no other source of macro data apart from the Federal Office of Statistics (FOS) is available for comparison. Part of the problem may lie at the FOS's estimate regarding the growth of output of the agricultural sector which is likely to be an underestimate for food crops (more on this in Chapter 6). In the same period, the GDP deflators rose by 12.4 per cent and 11.3 per cent for total and non-oil GDP, respectively, indicating a real GDP growth of 3.2 per cent ($R^2 = 0.79$) and a real non-oil GDP growth of 4.3 per cent per year ($R^2 = 0.58$) (Table 5.3 below).

Table 5.3 Percentage Annual changes of prices and output of main sectors 1973–81

	% change in current price	% change in constant price	% change in GDP deflator
Agriculture	12.4 (0.89)	−2.0 (0.67)	14.4
Manufacturing and mines	19.3 (0.78)	9.4 (0.61)	9.9
Construction	15.7 (0.90)	7.0 (0.82)	8.7
Services	17.0 (0.94)	5.3 (0.80)	11.7
(Trade)	19.0 (0.91)	4.4 (0.88)	14.6
(Transport)	19.4 (0.89)	5.2 (0.85)	14.2
Total (excl. oil)	15.7 (0.92)	4.3 (0.58)	11.3
Total (incl. oil)	15.6 (0.90)	3.2 (0.79)	12.4

Source: Calculated from the Federal Office of Statistics, *Nigerian Gross Domestic Product and Allied Macro-Aggregates*, Lagos, 1982.
Note: Figures in parentheses are coefficients of determination (R^2)

Using the consumer price index for the same period with an average growth of 16 per cent per year (IMF 1983) indicates little growth in the non-oil components of GDP, which seems rather implausible. Consequently, for the comparisons of the relative price and output growth in the following discussion, we have used the GDP deflator rather than the consumer price index, since the GDP deflator covers the entire economy whereas the CPI includes a more selective basket of consumer goods. A weighted average of nominal wage growth indicates that nominal wages rose at 15 per cent compound per year (Table B.11) which is above the average growth of the non-oil GDP deflator. The growth of nominal wages, however, is practically the same as the GDP deflator for agriculture, trade, and transport. Using the CPI, real wages would appear to have fallen by an average of 1 per cent per year in this period.

In the same period real agricultural output declined by 2 per cent per year, while the GDP deflator for agriculture rose by 14.4 per cent per year. This decline occurred along with the growth in the agricultural price index which is slightly above the rise in the GDP deflator for the economy as a whole. Changes in the agricultural output differ greatly year by year. For example, in 1974, agricultural output grew by 10.6 per cent and agricultural prices grew by 36 per cent per year, while in the following two years, although agricultural prices rose by more than the average inflation rate in the economy, agricultural output declined in absolute terms. A brief recovery in 1977 was again followed by a continuous decline of agricultural output accompanied by falling prices. Moreover, nominal wages in agriculture rose by an average of 20 per cent per year between 1973 and 1980, or 14.5 per cent if we take the 1972–9 period (Table B.9). Real wages in agriculture only rose in the first two years after the oil boom, declined until 1979, and then suddenly rose again in 1980. The contribution of agriculture to the GDP at constant prices declined from 31 per cent in 1973, to 22 per cent in 1981 (Table B.10).

Within agriculture, crop production showed the worst decline of 3 per cent per year, although food crops fared better than export crops. The only export crop showing positive production growth was palm oil (Table B.15). Livestock and fishing, however, grew gradually at a rate of 1 per cent per year. We shall discuss the agricultural sector in more detail later in Chapter 6. It appears that the national accounts are based on the Federal Office of Statistics' data which contain the most pessimistic estimates, particularly as far as food crops are concerned. Using alternative sources such as the USDA's data, for example, agricultural output would not have fallen as much as indicated by the national accounts. More on this in Chapter 6.

Manufacturing as a whole grew at a rate of 9.4 per cent per year in real terms while manufacturing prices grew at a rate of 9.9 per cent per year, hence lagging behind the average rate of inflation in the economy (Table B.9). Also, the growth of manufacturing was highly uneven. Except in 1977, the growth of manufacturing prices was below the average inflation in the economy while manufacturing output grew continuously until 1977, after which it went into continuous decline. The share of manufacturing and mining in GDP rose from 6.1 per cent in 1973 to 11.4 per cent in 1981. Nominal wages in the manufacturing sector rose on average by 15 per cent per year between 1973 and 1980, or 18 per cent per year if we take the 1973–9 period. Real wages in the manufacturing sector, however, declined continuously in the 1970s (Table B.11).

Within manufacturing, the output of both small- and large-scale manufacture rose by about 10.5 per cent per year (Table B.13). Between 1970 and 1978, the output of vegetable-oil processing, tin-smelting, and radio and television assembly declined, while there was little change in rubber and footwear manufacturing. Synthetic fibres grew most quickly, and vehicle assembly, soft drinks, soap, detergents, pharmaceuticals, beer and paints expanded at higher than average rates (Kirk-Greene and Rimmer 1981, p. 97; see also Table B.12).

The availability of cheap imported vegetable oil and the high cost of the domestic oil seed-crushing industry has resulted in the decline of such activities. Interestingly, the textile industry in Nigeria appears to have expanded rapidly after the oil boom, while in Iran, as we saw, the textile industry went through a difficult period of adjustment and concentration since only the capital-intensive industries could survive. The growth of vehicle assembly and textiles in Nigeria is partly the result of high levels of protection granted to these sectors. In spite of the high demand and locational advantages provided by high transport costs, the growth of the cement industry has remained sluggish (Forrest 1982, p. 332).

Construction and building growth rates were highly unstable (Table B.9). Rapid price rises in 1974 and 1975 were followed by sudden output growth in 1976 and 1977, followed by a period of recession in 1978 and 1979 which was again reversed in 1980. Large price rises in 1974 and 1975 in the construction sector were followed by a period of continuous real price decline from 1977 onwards. The output of the sector grew on average by 7.0 per cent per year and prices by 8.7 per cent per year, which is below the GDP deflator for the economy as a whole (Table B.8). The share of the construction and building sector in GDP rose from 10 per cent in 1973 to 12 per cent in 1981.

The price index of the trade sector rose by 14.6 per cent per year, which is above the average inflation rate for the economy, while the prices of services as a whole grew almost at par with the growth of the GDP deflator (Table B.8). The service sector is the only sector of the Nigerian economy which appears to have shown continuous growth, with the exception of 1978 (Table B.9). It should be added that, given the fact that many of the people in the service sector are self-employed, the rise in the relative price of services also implies a rise in the imputed wage of those engaged in services and they may have been the only sector with a rise in their real earnings in the 1970s. Government services are also part of the service sector and the Udoji awards of 1975 almost doubled the salaries of the civil servants. However, the inflation which followed these awards gradually eroded the real wage gains, so much so that the real wages (or salaries) of the civil servants in 1980 were below those of 1973 (Table B.11).

The wholesale and retail trade's contribution to the GDP rose from 19 per cent in 1973 to 22 per cent in 1981, almost reaching the importance of the agricultural sector (Table B.10). The service sector as a whole increased its share from 31 per cent of the GDP in 1973 to 37 per cent in 1981.

The picture that emerges for Nigeria is a complex one. An important point to note is the extreme volatility of price and output growth in the various sectors of the economy.

For 1973–81 as a whole, we find that the GDP deflator for trade, transport, and agriculture rose above the average GDP deflator for the economy, while that for manufacturing and, surprisingly, construction rose less than the average GDP deflator. The growth of nominal wages, however, is practically the same as the growth of the GDP deflator for agriculture, trade and transport. In other words, wages were stagnant relative to these three sectors while they rose relative to manufacturing and construction. In terms of sectoral price rises and the relative real wage, the Nigerian scene is different from the Iranian. In Iran, agricultural prices rose below the non-oil GDP deflator, while in Nigeria they rose below the average rate of inflation. Also the transport and trade sectors had contrasting experiences, with their prices rising below inflation in Iran and rising above inflation in Nigeria. In both cases, however, prices in the manufacturing sector rose below the average rate of inflation. Therefore, in terms of the inequality presented in Chapter 1, in Nigeria we observe:

$$\tilde{P}_a = \tilde{P}_s = \tilde{W} > \tilde{P}_m > \tilde{P}_c$$

In other words, in Nigeria prices in agriculture and services rose parallel with nominal wages while wages rose faster than prices in the manufacturing and the construction sectors.

Regarding wages, there is also an interesting contrast between Iran and Nigeria. Real wages rose in Iran relative to prices in the agriculture, manufacturing and service sectors, while falling relative to construction. In Nigeria the real wage rise was only clear relative to manufacturing and construction while wages did not rise relative to agriculture and services.

This again shows the importance of government policy in determining the outcome of the adjustment. Real wages rose more in Iran than in Nigeria and this is an important difference between the two countries. In Iran, the government followed an effective cheap food policy and the prices of imported staple foods, such as wheat, declined in real terms in the 1970s. In Nigeria on the other hand, the index of agricultural prices generally rose above the average inflation rate in the economy, and also food prices as a whole rose faster than the consumer price index (Table B.14). Moreover, comparing the previously mentioned measure of real appreciation, which is to adjust the exchange rate by the difference in the rate of inflation between trading countries, we find that in Nigeria the real exchange rate appreciated more than twice as much as it did in Iran. Table B.14 presents the results. Rising food prices in Nigeria and the consequent rise in the consumer price index occurred to a greater extent in Nigeria than in Iran (Table A.7). This is an important result which shows the key significance of government specific-sectoral policies (in this case food-pricing policy) in determining the outcome of the higher levels of expenditure for the economy and the intensity of the appreciation of the real exchange rate. A key result which emerges at this point is the predominantly non-traded character of food in Nigeria as compared to Iran, where massive food imports kept food prices below the rate of inflation. Food imports were facilitated not only by the financial strength of the government but also by the traded character of the main staple food in Iran.

On the output side, services grew at an average rate of 5.4 per cent per year; construction prices grew apace at first and then declined, while construction output grew on average at a rate of 5.8 per cent per year. The growth of construction is extremely unstable and the fact that the sector grew at an average rate of 7 per cent in real terms, in spite of the fact that wages were growing at double that rate, is somewhat puzzling. Government construction must have had an important part to play in raising the rate of growth of the construction sector, although data are not available on this point. The growth of the manufacturing sector, as we also mentioned in the case of Iran, is not fully compatible with the model unless we introduce government intervention in the form of tariffs and/or subsidies to the traded manufacturing sector, as has in fact been provided in both countries. One general point regarding sectoral GDP growth in Nigeria is the puzzling fact that the sectors with the lowest price rises showed the fastest real growth, namely manufacturing and construction, while sectors with the fastest price rises experienced the lowest growth rate, namely agriculture and services. In other words, in terms of explaining real output growth, the model presented in Chapter 1 does not allow very definite results. Nevertheless, by introducing the role of government, one can obtain more definite results which, though modifying the predictions of the model, maintain its overall strength in explaining economic performance in these two countries.

To see what other factors could explain sectoral performance, Chapter 6 takes a closer look at Nigerian agriculture.

Notes

1. Looking at the capital account, it is clear that 'non-adjustment' or capital outflow has not been significant in Nigeria. Capital (other than reserves) has been positive in all years except 1974 (where it was negative to the tune of $70 million) and 1976 ($67 million). Calculated from IMF (1983).
2. Calculated from ibid.
3. The derivation of the domestic deficit is from the 1981 country economic memorandum on Nigeria by the World Bank. This publication derives the deficit by using specific assumptions concerning the share of foreign payments in total government expenditure. More specifically, it assumes that the public sector pays 20 per cent of its current expenditure and 5 per cent of its capital expenditure abroad.
4. There are a number of studies on the long-run income elasticity of demand for money in Nigeria. Ajayi (1978) estimated the elasticity at about 1.1, Morgan (1979) at 1.4, and Teriba at 1.5 for the periods 1960–75, 1961–76, 1958–72 respectively. The equations are all in log-linear form using real income as the scale variable and either the rate of interest or the expected price level as the opportunity-cost variable. The opportunity-cost variable is usually insignificant and therefore the real demand for money is viewed as a function of real output growth.

5. In 1980, total reserves minus gold stood at 10.2 billion naira and imports were 11.8 billion naira in 1983, total reserves minus gold stood at 963 million naira, and imports in 1982 were 10.8 billion naira. See IMF (1984, pp. 453, 455).

6 Nigerian agriculture in the 1970s

Before analysing the production performance of a number of important crops in Nigeria, it is necessary to discuss the socioeconomic environment of the agricultural sector since, as in the case of Iran, a number of institutional and socioeconomic factors play a major role in explaining the supply response to price changes. In other words, price and exchange-rate changes are only among a number of factors which affect agricultural production. These factors include the land-tenure system and farm size, the farming systems and labour requirements, the extent of commercialisation of agriculture, and government input and investment policies, including the credit policies towards the agricultural sector. After this background is clarified, we will proceed to discuss some of the important food and export crops in the light of the previous discussion and by considering relative-price changes facing the crops. In this way, supply response will be placed within a broader and more comprehensive framework, and it will be shown how terms-of-trade changes are by themselves insufficient in explaining supply response. Also, the broader socioeconomic framework sheds further light on the theoretical framework presented in Chapter 1.

The gist of the argument concerning institutional, policy, and price factors developed in the next section is as follows: the majority of Nigerian farmers are small-scale operators with owner-like possessions of the land. Both the size of farm and the types of title to land are problematic as far as access to credit and farming improvements is concerned. Moreover, the rising opportunity cost of labour, the stagnation of prices of export crops, competition with imports, and the rising price of food have meant a shift of labour and resources by the small farmers towards the production of food crops. In addition, within food crops, the high-value irrigated cash crops such as vegetables, rice, maize and sugar-cane have become quite profitable and have expanded more than the traditional food crops which have been relatively stagnant. Export crops have shown a general decline.

Moreover, this trend has been apparent at least until 1978 since between 1978 and 1983 rural wages nearly quadrupled while food prices do not appear to have risen as fast. Wages rose due to the sharp acceleration of public expenditure after the second oil price rise in 1979 with the consequent increase in demand for labour and increased levels of food imports. Hence, except for a small proportion of mechanised farms, the bulk of food producers involved in the market appear to have faced a difficult economic environment. This points to a severe labour constraint facing the agricultural sector as a whole, due to greater off-farm and urban opportunities, which has resulted in the slow growth of agriculture and the decline of its contribution to the GDP. Government policy, on the other hand, has been bimodal with a bias towards large commercial and state farms in terms of investments, credit and inputs. The small-farm sector has been dynamic in particular sub-sectors in spite of government policy and the labour constraint. This double squeeze has had a high opportunity cost for the economy since it has deprived a very large, resource-based and active sector of resources. In sum, what is observed in the agricultural sector is its shrinking size in terms of its relative contribution to the GDP; a shift towards food crops in general and high-value food crops in particular, especially by the small farmers; and a policy bias in favour of large, capital-intensive farming which is largely self-defeating since this is not where the greatest potential for agricultural development lies. Again, the conclusions broadly confirm the theoretical framework of Chapter 1 and highlight the importance of government policy in affecting the outcome of the adjustment process, both at the macroeconomic and at the sectoral level.

The land tenure system

The Nigerian agrarian structure is characterised by the prevalence of small farms. Between 6 and 9 million families, with an average cultivable land of 0.9–2.5 hectares, are the majority of farmers in Nigeria.[1] Table 6.1 shows the farm size, the number of households, and the percentage of the farmed area in each category for 1973–74. Over 80 per cent of the farming households operate farms of less than 2.5 ha and this constitutes 65 per cent of the total farmed area. Moreover, the bulk of the operators have owner-like possessions of land which is either inherited or allocated by the family or village head, and is often made up of two or three scattered plots around the village. In 1977–78, 60 per cent of all farms were inherited, 30 per cent were allocated by the family head, 5 per cent were purchased, 4 per cent were rented, and 1 per cent were communal or squatters, Federal Office of Statistics (FOS), 1978–9, p. 5.

Table 6.1 Land distribution, 1973–74

Farm Size (ha)	% of households	% of farmed area
0.00 - 0.25	16	1
0.25 - 0.50	20	3
0.50 - 1.00	16	15
1.00 - 2.50	29	47
2.50 - 5.00	18	26
Over 5.00	1	8

Source: Calculated from Federal Office of Statistics, Rural Economic Survey, 1973–74.

Tenure arrangements and the form of access to land by the small farmers are quite varied in rural Nigeria. The two main customary forms of land tenure are *gandu*, under which the land right is vested in the family head but the land is worked and its produce shared by all the family members, and *gayauna*, under which the land is worked by the family member who has the right to use the land. The *gandu* is of particular importance as it enables the individual to hold off-farm employment while still enjoying the benefits of agriculture. The landholding groups can also be the clan, the village, or the community. The member of the relevant unit is entitled to a block of land and is restricted from transferring ownership, mortgaging, or selling his land at will, without the consent of the relevant social unit and, since 1978, without the permission of the local government because of the land use decree. Until the 1970s, customary land-tenure remained the principal form of tenure throughout Nigeria. During the decade, as land prices rose sharply, communal tenure gradually gave away to individual ownership, or to multi-year tenancy rights (Agbola 1979, p. 36). Nevertheless, absence of legal title to much of the land has hindered the purchase of land. This is clear from the small share of purchased land in total-tenure since, as was referred to above in 1977-8, only 5 per cent of all farmed land was purchased. Ike (1977, pp. 88–9) has shown, for example, that the income of farmers with clearly defined property rights to land was 20–30 per cent higher than the income of communal tenants.

The ambiguity concerning rights to land is an important factor which has negatively affected output and investment in Nigerian agriculture. The small farmers are both subsistence and market-orientated and security considerations are the critical determinants of the extent of their involvement in the market. Farmers are, however, handicapped by their poor education, low income, and considerable difficulty in obtaining credit. Furthermore, they usually live in neglected rural areas with poor physical infrastructure

(Idachabe *et al.* 1981).

Apart from the small-holder owner-possessor-occupier, there is also a group of migrant-tenant farmers. The migrant-tenant farmer operated 4 per cent of the total farmed area in 1977–8. There are usually young adults, aged between 25 and 44, who left their home district because of the shortage of farmland. Migrant-tenant farmers usually plan to return home with capital to set up a business or learn a profession. They generally settle in easily accessible rural areas so that their produce can be readily marketed (Udo 1975).

Apart from small-holders (owner-occupiers, tenants), a small but increasing number of large capitalist farmers constitute a third group of producers with farms of over 5 ha and usually ranging between 20 and 500 ha. These farms are sometimes inherited but more often are owned by senior civil servants, army generals, wealthy politicians, and directors of private companies. Many of these elite farmers are well educated and acquired the land they now cultivate when they held influential positions in the public sector. They are generally creditworthy and have ready access to senior government officials (Udo 1983, p.11). This group of farmers produces entirely for the market and especially for the urban markets. As profit-seeking entrepreneurial units, such large farms are generally well organised and financially quite successful; they are usually involved in poultry, maize and high-valued vegetables. It is this group of farmers that appears to have benefited most from the government provision of credit, services, subsidised inputs, and guaranteed prices–the so-called 'green revolution' programme which is discussed later in this chapter.

Finally, there is another group of producers who are best referred to as 'backyard gardeners', mostly consisting of the enterprising public servants and innovative households who produce food either as a hobby or, more likely, in order to reduce their expenditure on food – the Operation Feed the Nation Programme (discussed later) of 1976-9 was primarily directed towards this category of part-time farmers (Udo 1983, p.12).

Before leaving the discussion of the forms of land tenure, another development should also be discussed and that is the land-use decree of 1978, already referred to above. The land-use decree vested all lands in the state in terms of ultimate control and formal ownership, while confirming existing occupancy rights, but with some important changes: first, it instituted individual tenure instead of communal tenure, that is, occupancy in the name of a single person instead of families or lineages; second, it prohibited the transfer, mortgage or sub-lease of land without prior consent of the local government; third, it prohibited fragmentation of land without the prior consent of the local government; fourth, it made occupancy rights revokable if the land in question was required for public purposes (Sano 1983). Further, the decree gave state and local governments paramount authority to take over any undeveloped land and assign and lease it as required. In effect, the decree nationalised all land by requiring certificates of occupancy from the government for land held under customary and statutory rights, and the payment of the rent to the government.

The impact of the land-use decree on the smallholders appears to have been detrimental. The decree has in fact attempted to achieve contradictory aims. By confirming individual instead of communal tenure, the decree appears to have facilitated greater individual incentives; nevertheless, by reducing all land users to lease-holders and stating the authority of the government over all land, it may have increased insecurity of tenure and lowered the incentive to improve land quality since all land is subject to confiscation by the government. So far there has been a number of incidents of resistance and rebellion by the peasants to the takeover of land by the government for large-scale irrigation projects at Sokoto, Bakolori, and the Tiga Dam (see pp. 134-7)

Farming systems and constraints

Although the term 'Farming System' has quite a broad connotation in the development literature, here it is used to refer to the rather narrow aspect of farming, namely the crop mix in terms of the combination of food and cash (or export) crops by the farmer.[2] After discussing the different agricultural zones and the crop mix, some of the constraints, especially labour and the credit constraints facing small farmers, are discussed. Furthermore, policy factors with implications for the farming systems are explored on pp.133–40 although some references to credit policies are also made in this section in the discussion of the credit constraint.

Crop mix

Broadly, one can distinguish two main types of farming system in Nigeria. These are the root-crop-based farming systems in the south and the grain-based farming systems in the north. There is also an intermediate system, as one moves northwards in the mixed-crop zones of the middle belt where the climatic conditions allow the growth of both root and grain crops Figure 6.1. The two main systems are quite labour-intensive but they differ in the kinds of crop and crop mix. The two systems, among other things, reflect the variations in the length of the growing period, which is gradually shortened as one moves northward.

In the south, the farming systems are largely based on yams and/or cassava as the main food crops while export crops are palm-oil in the south-east and cocoa and rubber in the south and south west. As one moves north, sorghum gradually acquires importance. In the rural areas of Nigeria especially in the south, yams account for 20 per cent of the daily calorific intake, although there are sharp seasonal variations (Diehl 1982, p. 3). The importance of yams in the south is surpassed by cassava only in the southern forest belt. These two root

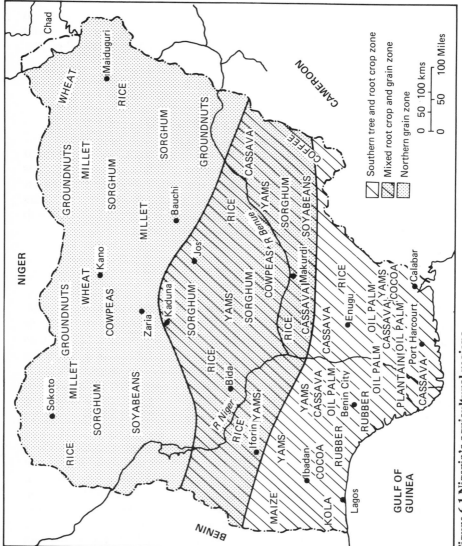

Figure 6.1 Nigeria's agricultural regions

crops constitute the major part of the diet in the southern part of Nigeria. In terms of the contribution of yams to farmers' income, various studies have shown that it is about 30 per cent of the farmers' gross income (Lagemann 1977, p. 228). Yams or cassava are, of course, only the basis of the farm system, while the whole system is composed of a number of tree and arable crops. A multi-storey physiognomy, with perennial and annual crops, is typical of the village compounds in eastern Nigeria. The important tree crops are oil palms, coconuts, kola nuts, mangoes, oranges and bananas, while the important arable crops are yams, maize, cassava, cocoyams, beans, pepper and vegetables. Various trees form the upper part of the storey, whereas arable crops grow under the shade of the trees (Lagemann 1977, p. 32). As one moves westwards the importance of palm oil decreases, and first rubber and then cocoa become important.

Moving north towards the middle belt, the importance of export crops such as palm oil decreases while sesame acquires importance and there is an increase in the cultivation of food crops, especially rice and sorghum. North of the middle belt lies an extensive area dominated by cereals where, in contrast to the south, a short wet season, with light rain fall and low humidity, characterises the climate. Principal food crops are millet, sorghum, rice (upland and irrigated), beans, cowpeas, groundnuts (also an export crop), and irrigated cane-sugar. The principal export/cash-crops of the north are groundnuts, cotton and tobacco. Millet is the dominant crop in the whole of the north. Wheat is also grown in small quantities around Kano and Maiduguri. In Kano state, there is also permanent cultivation of market-garden crops.

A relatively small but important farming practice, especially in the north, is the cultivation of the lowland flood plains known as *fadama*. The term is used by the farmers in the Nigerian savannah to distinguish the valley bottom soils or *fadama* from the upland soils, referred to as the *gona* (Norman et al, 1982). Traditionally, the *fadama* areas are used in two ways both for cash income and for food. During the wet season, low-lying river-valley fields which are temporarily inundated, are planted with rice, sugar-cane and tobacco with receding flood; in the dry season, the *fadama* soils are used for growing vegetables such as onions and tomatoes, with water lifted from hand-dug wells or riverstreams by *shadoof*. Recently pumps have also been used. The total net area of *fadama* with small-scale water control is estimated by the World Bank (1979, Annex IV) to be around 800,000 ha and the area appears to have grown rapidly in the 1970s.

Some important constraints

Having described the main crop mixes in the different agricultural zones, we now consider some of the other constraints apart from the land-tenure issue which have affected the performance of the farming systems in Nigeria. This background, combined with a discussion of government policies directed towards agriculture, is necessary for a better understanding of Nigeria's crop level performance in the 1970s, which is discussed later in the chapter. The key issues discussed here are those of labour and credit, and also, more briefly, the marketing issue.

Labour constraint. Labour is the principal input into agricultural production in Nigeria and mechanisation is largely confined to large public sector projects in food production which are still insignificant in terms of their contribution to national food supplies. The rising opportunity cost of labour due to the availability of off-farm occupations, the Universal Primary Education Campaign, and the drudgery of low-productivity farming have all contributed to creating an acute labour shortage and rising labour costs in Nigeria's agricultural sector. Hired labour costs vary from 23 per cent of the total value of crop sales in Bendel state to as high as 67 per cent in Imo state, according to the APMEPU Baseline Survey conducted in 1980 (World Bank 1984d).

The index of nominal wages for the agricultural sector stood at 268.5 in 1978 and 522.5 in 1980 (1973=100). This meant that daily agricultural wages rose from 0.70 naira to about 2 naira in the 1973–78 period. In real terms, agricultural wages were largely static in this period since food prices rose at more or less the same rate. The price index of Nigeria's major export commodities (excepting groundnuts and palm oil which became domestic food crops in this period), however, had rises nowhere near that of the agricultural wage rate (Table 6.2).

Table 6.2 Index of agricultural wages and prices of major export crops, 1980 (1973 = 100)

Agricultural wages	522.5
Benniseed	300.0
Cocoa	367.0
Cotton	303.0
Groundnuts	587.5
Palm kernels	295.0
Palm oil	535.7
Soyabeans	319.1

Source: Calculated from Tables B.11 and B.26

It follows that only food producers were able to afford hired labour and export producers' margins must have been sharply reduced. Also, there is no evidence of government subsidies to the export producers, a point which is discussed further in the next section. A sharp rise in agricultural wages occurred between 1978 and 1983 when daily rates for unskilled work in peak periods during the crop season jumped from 2 naira to nearly 7 naira in the south (World Bank 1984a, Vol. 1, p. 7). In the same period, the consumer price index rose from 80 to 156 (IMF 1984, p. 455), hence the real agricultural wage rise in the 1978–83 period was considerable three to five times higher than the wage rise in competing LDCs measured at official exchange rate (IMF 1984, p.8). The 1979 second rise in oil prices, and the massive injection of public funds into the economy for largely infrastructural and construction projects, undoubedly were a major force behind the sharp rise in rural wages in the 1978-83 period. Another important factor was universal primary education, which became effective from 1976. Enrolment in primary schools rose from 5.9 million pupils in 1975–6 to 12.7 million in 1979-80 and the target for enrolment in the Fourth Plan was 17.5 million pupils by 1984 (NPO 1981, p. 265)

To cope with this massive rise in enrolment, increasing resort has been made to unqualified teachers, which indicates that the cost of the production loss to agriculture through the drain of the younger generation may well be far above the benefits of universal primary education to the national economy. This is a critical point concerning the government's development strategy, though not one which can be discussed at any length here.

The case studies carried out in southern Nigeria showed that the smallholders surveyed employed about 30 per cent of farm labour outside the farm family. The data indicate that farmers in areas with high population densities employed more outside labour to compensate for the relatively high incidence of off-farm activities. In response to the question of why farmers did not increase their size of farm, almost 32 per cent gave labour shortage as the answer, and 29 per cent gave the shortage of land as the main cause (Okafor 1975, pp. 114–116).

In Hausaland in northern Nigeria, which has a lower degree of commercialisation of agriculture and where there is no permanent class of landless agricultural labourers, hired labour contributed to 16 per cent of the total labour input on the family farms in 1979 (Norman *et al.* 1979, p. 33).

The contrast between south and north is interesting since greater dependence of agricultural production in the south upon hired labour implies that the rise in agricultural wages must have exerted greater direct pressure on farmers in the southern part of Nigeria. The southern rubber and palm-oil plantations appear to have suffered considerably from this relative-price development, as is discussed later.

What emerges from the above discussion of labour cost and availability is that there was a general rise in cost and a fall in the availability of agricultural labour. Until 1978 however, food producers managed to keep afloat because of the simultaneous rise in the rising cost of labour and the rise in food prices. Export-crop producers must have fared much worse since the prices of export crops facing the producers did not rise as fast. The relative price parity of domestic food prices and wages was broken in the 1978–83 period, with wages outpacing prices at a rapid rate. It should be added that this argument is consistent with the earlier-stated view concerning the profitability of a number of non-traded food crops, at least until 1979, since until then food prices were rising along with wages, but this trend no longer appears to be the case.

Credit constraint. Increasing commercialisation and increasing reliance on hired labour have brought the credit constraint to the forefront of discussions on Nigerian agriculture. Moreover, the crucial bottleneck is for working capital credit to pay for the increasing reliance on hired labour. The main sources of rural credit in Nigeria are relatives, friends, informal credit societies, the village money-lenders, and the formal credit institutions which are the commercial banks or the state government credit agencies.

The bulk of credit used by the small farmers comes from the informal sector (Ngozi 1982, Part VI). In a survey of farmers' indebtedness to the informal sector, it was found that in Nigeria as a whole 45 per cent of loans were received from relatives, 30 per cent from friends, 20 per cent from informal societies, 3 per cent from patrons, and about 2 per cent from the money lenders. Concerning the use of credit received by farmers, 62 per cent has been directed to non-farm use and 38 per cent to direct farm use. The bulk of non-farm credit has been used for school fees and housing. Under farm-use, the requirements for paying hired labour accounts for 75.7 per cent. This is especially significant since it demonstrates the importance of working capital credit in the total on-farm credit requirements of the farmers. This credit constraint becomes particularly binding under the situation of rising labour costs which has been the case in Nigeria in the past decade.

The informal credit societies are largely concentrated in the south. Also there was very little evidence of loans from professional money lenders in the north. The information concerning the informal money-lenders is rather scarce. The rise of village money-lenders appears to be of recent origin (Awa 1973, p. 9). Only one loan out of 73 in the Ngozi's sample came from a money-lender. Their interest rates are quite high. The one money-lender loan in the sample bore a 60 per cent nominal per annum rate compared to the mean rates of 17 per cent on loans from relatives, 34 per cent from friends, and 29 per cent from informal credit societies. Also, contracts with the money-lenders are very rigid. Defaulters often forfeit title to inherited land against which the loan has been secured, usually with the concurrence of their kin. Farmers are therefore generally reluctant to approach the professional money-lenders for credit (Awa 1971, p.9).

The formal credit institutions have generally been much better studied. The first point to note is that agriculture and related activities have generally received a small share of commercial loans and public lendings. Between 1970 and 1977, the share of agriculture in total commercial bank loans never exceeded 4.5 per cent (Osuntogun 1983, p.10).

Also, only 8 per cent of the already small share of agriculture in the Fourth National Plan is allocated to a number of items which include credit (World

Bank 1974, p. 57). The Nigerian Agricultural and Co-operative Bank (NACB) is the public sector source of credit to the agricultural sector. In the 1974–80 period, a total of 241 enterprises were assisted, the bulk of which are large-scale farms or co-operatives (Osuntogun 1983, p. 9). The domination of village co-operatives by the traditional leaders and the local elite and the accrual of the bulk of co-operative credit to the more prosperous and influential farmers have been documented by King (1981, p. 278). NCAB's loans have not benefited the small family farmer and the main source of credit remains the family and friends – this is hardly a reliable source in times of a severe working capital bottleneck due to rising wages. In order to liberalise the lending policy of the NACB, a *Small Holder Direct Loan Scheme* was started in 1981 whereby the federal government guaranteed the loans of the peasants who were not able to present the security required by the NCAB, so that they could now approach the bank without the prior consent of local state authorities (Sano 1983). Although it is too early to determine the impact of the direct lending scheme, it is difficult to see how it would make much difference to the credit situation of the small farmer, given the complicated institutional issues surrounding the problem of small-farmer credit.

In 1977, the government passed a decree establishing the Agricultural Credit Guarantee Fund which would guarantee loans of up to 75 per cent of the value of the principal and interest for loans granted by commercial banks to the agricultural sector. Also, in the same scheme, the government extended a graduated tax-free allowance to commercial banks on interest earned from agricultural loans depending on the duration of loans – the longer the duration the greater the tax holiday (Osuntogun 1983, p. 9). Under this scheme, the banks were required to increase their number of branches by 260 or more by the end of 1983.

Looking at the sectoral distribution of credit in Nigeria, one notices a progressive rise in the share of agricultural credit in the total commercial banks' credit portolio. In 1972, agriculture received 2.9 per cent of all comm-ercial bank credit. The share of agriculture rose to 4.5 per cent in 1977, 5.5 per cent in 1978 (Okigbo 1981, p. 62) and 7.2 per cent in 1981 (Central Bank of Nigeria 1983a, p.57). Although the share of agricultural credit in the sectoral distribution of commercial bank credit has risen, there is a problem of interpre-ting what these figures mean in practice. Under pressure from the Central Bank to expand their agricultural lending activities, the banks apparently applied a very broad definition of the agricultural sector, and the greater the pressure on them to lend to this sector, the broader the definition became. Hence, processing ventures, forest industries, and even finance for food dis-tribution could be and were slotted into this portfolio (World Bank 1984a, p. 28).

Moreover, the problem of access to these agricultural loans by the small farmers who are the backbone of Nigerian agriculture remains highly problematic. Commercial bank branches, as well as the offices of NACB, are concentrated in the urban areas and are inaccessible to most rural folk. Moreover, there are institutional obstacles to the alienability of land which constitutes the main asset in the traditional farming system. This has resulted in a general unacceptability of land as loan collateral by the formal credit institutions (Ngozi 1982, p.182). Other problems, such as the literacy requirement, fear of disciplinary action in case of default, the banking system's orientation towards the rich, and the small size of the loans are some of the other factors aggravating access to credit by the small farmers. A 1974-75 survey of small farmer credit in the south and southeast of Nigeria, showed that the absence of collateral and the fear of default were among the cheif reasons preventing small farmers from seeking credit from the formal sector. Other factors included inadequate knowledge or lack of knowledge concerning the existence of such credits (Awa 1971, p.10)

Marketing. Both the government and the private traders are active in marketing of farm produce. Government marketing is done through eight marketing boards which cover both export and food crops. The boards operate a minimum pricing system, both directly and through buying agents. The boards enjoy exclusive control over exports but not over the marketing of the domestic food crops. Prices are generally fixed by the Head of State on the advice of the Technical Committee on Producer Prices, an independent body chaired by the Minister of Finance and supported by the Central Bank which comes to agreement concerning farm-gate prices and other allowances that are provided to the licensed buying agents (LBAs). In general, food-marketing boards have been far less effective than the export-crop marketing boards. Moreover, since a number of export crops such as palm oil and groundnuts are now largely used as domestic food crops, a general weakening of the government marketing role has resulted since government prices are usually below the market price for such crops and an increasing proportion of these crops do not enter the official marketing channels (World Bank 1979, Vol. 2, p. 51).

Concerning private traders, the evidence suggests that marketing channels are relatively effective and do not reflect monopolistic or monopsonistic conditions. The existing weaknesses in the private marketing systems, therefore, do not arise from manipulations by individuals or the state but, rather, result from high risk and gaps in inter-market information. Erratic supply, for example, increases the risks associated with specialising in marketing. Traders tend to be locally orientated, resulting in the absence of a nationally integrated marketing system (Hags and McCoy 1978). In some cases, also, poor farmers may have to rely on traders for credit, selling produce to pay the loan when prices are at a post-harvest low. Having sold their produce early in the year, they may need to borrow for the next season, leaving them in a debt trap. Here private traders can use their credit power to pay farmers less than the market prices. If anything, this situation again points toward the credit constraint already discussed.

Concluding remarks. Labour and credit have been emphasised as the key constraints facing small farmers in Nigeria. We have underlined the interconnectedness of these two constraints, that is to say that the labour constraint is aggravated by the credit constraint while the credit constraint is in large part the result of the labour constraint. The crucial bottleneck is for working capital credit to pay for the increasing reliance on hired labour. Under farm use, as was mentioned earlier, the requirements for paying hired labour accounts for 75.7 per cent of the total credit. The credit constraint becomes particularly binding under the situation of rising labour costs, which has been the case in Nigeria in the past decade. Also, evidence has been brought forth to show the lack of access of small farmers to formal institutional credit. Finally, the reduced role of government in marketing food crops and the relative effectiveness of marketing channels imply that the earlier-stated constraints are the more important ones, especially for small farmers. Apart from this background on the farming systems and constraints, it is also necessary to discuss the impact of government agricultural policies in improving production conditions in Nigerian agriculture, in order to provide the background for a more detailed discussion of prices, production, and income at crop level.

Impact of agricultural policies

Higher oil revenues have had a distinct impact on the types of policies directed towards agriculture, hence policy is more of an endogenous rather an exogenous factor in this context. Two broad tendencies are observable in the types of policy introduced: increased direct government entrepreneurship in agricultural production; and the dominant role of the federal government in financing agricultural development. This enhanced role of the federal government in the agricultural sector is reflected in the three main strategies which have been formulated and applied in the 1970s and early 1980s:

1. Introduction of large complex irrigation projects under the direction of the newly founded River Basin Development Authorities (RBDAs). RBDAs have been increasingly absorbing the bulk of federal government agricultural expenditure since 1973 when the first one was created. The basic components of the river irigation projects are: state management, modern technologies, and the cultivation of import-substitute crops, especially wheat.
2. The World Bank-assisted Agricultural Development Programmes (ADPs). This strategy is an attempt to reach the smallholders. With the Bank's assistance, this project is intended to cover the whole of Nigeria.
3. The Green Revolution Strategy, aimed at raising rural productivity and food production through the distribution of subsidised credit fertiliser, improved seeds, tractors, and some changes in the tenure.

All the three strategies are attempts to increase food production. Export crops have become largely neglected in the new strategy. Moreover, the policies followed are based essentially on the belief that increased productivity can best be achieved with new (imported) technology, foreign expertise and management, and reliance on 'progressive' farmers.

Large-scale irrigation

The series of droughts that hit the Sahelian zones of northern Nigeria in the early 1970s, and rising imports of a number of key foodstuffs, such as wheat, sugar and rice, resulted in an increased emphasis on irrigation developments especially in the Third and Fourth Plans (1975–85). The new importance attached to irrigation, resulted in the creation of a separate Ministry of Water Resources and the establishment of 11 River Basin Development Authorities (RBDAs). The RBDAs have fairly broad responsibilities including the development of irrigation schemes, and schemes to develop water supply, flood control, pollution control, and resettlement. The federal Ministry of Water Resources, of course, is in charge of co-ordinating the activities of the RDBAs and giving policy directions. Moreover, the RBDAs are entirely financed by the federal budget.[3]

Since RBDAs are federally funded, the federal government has greater authority over their activities and this has resulted in tension between the state and federal governments. Existing irrigation projects under the authority of state governments have been transferred to the RBDAs. Also the RBDAs have attracted the technical staff of state governments.

Moreover, the whole centralised and federally controlled approach of the RBDAs has been criticised as unconstitutional since it infringes upon the authority of the state governments (Awojobi 1982).

In any case, a series of major irrigation schemes have been started since 1973 principally to produce wheat and rice and often with little initial preparation.[4] The idea was to prepare the land, build the irrigation infrastructure, and allocate plots of about 5 ha to smallholders who are tenants for cultivation of specified crops, namely rice, wheat and some vegetables. The projects would also provide support services to the farmer-tenants to improve their productivity. By 1979, a theoretical 66,000 ha were being developed in Sokoto, Kano and Borno states. These were the Bakolori Dam Project, the Kano I River Project, and the South Chad Irrigation Project. At the same time, an additional 167,000 ha were being studied. A government committee in 1979 declared that if all projects proceed according to plan, 165,000 ha would be under irrigation by 1983 (World Bank 1979). The actual irrigated area in 1981, however was about 13,000 ha (FAO 1983a, p. 5). Of the 13,000 ha, 8,800 ha was under rice and 4,900 ha was under wheat, with respective average yields of 2 tonnes and 3 tonnes per ha. The requested budget of the eleven RBDAs for 1982 was 500 million naira ($800 million) (FAO 1983a, p.5). This should be compared with an irrigated area of 20,000 ha, allowing for some double cropping and some increase of irrigated area between 1981 and 1982. The requested budget cost per hectare comes to about $40,000. In 1984–5 the total irrigated area under RBDA operation was 28,000 ha and the total allocated budget for 1981–5 was 1.8 billion naira ($2.5 billion). Taking five times 28,000 or 140,000 ha (as the gross irrigated area for the 1981–5 period), the cost per hectare appears to rise to almost $18,000 per year which is a phenomenal sum.[5]

These high costs are attributable to poor design, preparation and management. The so-called design and construct contractual agreements, which involve a built-in tendency to propose the most expensive designs, have been common. Also, costly civil construction work and delays have been another aspect of these schemes. Considerable delays have been experienced concerning the training of technical staff and poor management has been the result (World Bank 1979, p. 8).

Agronomic aspects have also been badly neglected. The most immediate concern is the effect of undisciplined water management leading to water logging, particularly where surface irrigation is carried out on light soils overlying shallow impermeable bedrock. This is the case for example with the Kano River Project, where waterlogging is already a problem although the project is only 25 per cent complete. Many parts of northern Nigeria have this sort of soil condition and extreme care has to be taken to provide for adequate sub-surface drainage (Carter *et al* 1983, p. 74). Moreover, the high percentage of light, textured and permeable soils, as in the Kano River and Bakolori schemes, reduces the scope for the cultivation of high-value crops like rice and wheat in favour of less demanding crops that can also be grown under rain-fed conditions, such as sorghum, groundnuts and cowpeas (World Bank 1979, p. 7).

Apart from the high costs and poor management, the approach of the RBDAs has been top-down and autocratic. To prepare the land sites, existing land-holders have been forcibly removed with insufficient compensation. According to Williams (1983)

> Compensation, in land or in cash, has never been adequate, timely or, often, even forthcoming. Many farmers deprived of the means to farm left to seek urban employment. On the Bakolori scheme in Sokoto state in the north, farmers refused to allow contractors to destroy standing crops. Others refused to accept the infertile land allocated to them. Others were unable to cultivate their lands for three seasons. Farmers demanding compensation obstructed contractors and disrupted the irrigation system until in 1980 the police attacked them, brutally killing hundreds of men, women and children.

Another source of conflict with the farmers has been over the cropping pattern. There is a conflict between growing wheat in the dry season and growing sorghum in the wet season since sorghum is harvested in December, while land preparation for wheat must begin in October if yields are not to drop significantly. In fact the farmers are instructed from above not to grow sorghum, but they continue to do so (Williams 1983; Wallace 1981, p. 243).

What the schemes have ignored is that for the farmer the need to secure the family's food supplies is of paramount importance in spite of the high returns which may be available from irrigated wheat farming. Farmers are being asked to become fully specialised in production for the market, and give up their traditionally more reliable sources of food supply. The price of foodstuffs, as is shown later in this chapter, has risen substantially. Given the unreliability of irrigated farming due to poor management and inadequate supply of inputs and support services, the farmers will be accepting too high a risk in following the direction of the schemes. A recent survey of major constraints in RBDA projects has shown that many farmers have defected from the schemes (FAO 1983a, p. 89).

In addition to the problems in the project areas, the irrigation projects have also affected farmers in the periphery of the schemes. Traditional *fadama* irrigation, referred to earlier in this chapter, has suffered badly from the large schemes since farmers have lost fertile flooded land downstream. In Sokoto state, farmers lost over 80,000 ha of *fadama* lands (Palmer-Jones 1984). The irrigated area of the Kano River Project I in 1984 was 14,000 ha (FAO 1985). Stock (1977) estimates that the impact of the decline of the floods due to the Tiga Dam on the river Kano has been the loss of 75 per cent of rice production from the *fadama* land downstream. It should be noted that the population in the Hadejia valley alone is around 250,000, while the number of farmers working on the Kano River Project does not exceed 10,000. The loss of existing small-scale irrigation also indicates that net irrigated hectares in the official schemes cost even more than the data on p.135 for gross irrigated hectares suggest

The evidence overwhelmingly shows that the large-scale irrigation projects which have been a corner-stone of the Third and Fourth Plans, have not been successful. This is a clear case which shows that it is not only how much is spent that affects the outcome of adjustment to higher oil revenues but that how the money is spent is of crucial importance. Of course, one cannot argue with the need to have effective water management in the drought areas but the question is what are the best means of achieving water control given the existing social, economic, and managerial constraints present. The obvious alternative to the large-scale irrigation projects is improvement and expansion of the existing small-scale *fadama* irrigation which has evolved to fit the local farming systems. In fact, the World Bank assisted Agricultural Development Programmes have put considerable emphasis on the potentials of smallholder irrigation, but the volume of resources allocated to this category is quite small in comparison to the resources allocated to large-scale irrigation.

Several studies have shown the highly profitable nature of *fadama* farming which should now be discussed. Reference has already been made to the extent and nature of *fadama* farming in northern Nigeria. There are indications that smallholder development of *fadama* has been expending at up to 20,000 ha per year from 100,000 ha in 1958, to about 800,000 ha presently (FAO 1985, p. 9). Farmers either rely on the residual moisture, or the *shadoof*, which consists of a bucket and a long wooden stick to take water from the river; also recently pumps have increasingly been used. There has been a booming market in water pumps: during 1983 and 1984 over 10,000 pumps were sold in Bauchi and Kano state (Fadama Survey Report 1984, p. 12).

Studies of costs and receipts reveal the highly profitable nature of the *fadama*. Total costs per hectare of three different types of *fadama* in Bauchi and Kano states using residual moisture, *shadoof,* and water pumps were 853, 1,263, and 1,506 naira per hectare, respectively, during the dry season for vegetable cultivation. Total receipts for the same farms were 1,464, 4,757, and 5,018 naira per hectare, respectively. The gross margins, therefore, were 611 naira per hectare for the residual moisture method, 3,494 naira per hectare for the *shadoof* method, and 3,512 naira per hectare for the pump method (Fadama Survey Report 1984, p.15). No studies have been available to the author concerning the profitability of rice-farming on the *fadama* during the flood season but the rapid growth of irrigated rice farming indicates its highly profitable nature. Moreover, an important advantage of the *fadama* lies in their complementarity with upland agriculture where the farmers produce their own staple food requirements. The low cost of irrigation in the *fadama* should indeed be compared to the astronomical sums involved in large-scale projects. Of course, at present, *fadama* land is increasingly scarce but investment in this type of irrigation can bring more land under this sort of cultivation. As was mentioned earlier, the government appears to have taken land off the *fadama* cultivation due to its large projects. Sano (1983, p. 39) estimates that, in 1980, 70 per cent of the total annual expenditure on agriculture went into large irrigation projects while the Agricultural Development Programmes, with their important *fadama* component, received only 4 per cent of total agricultural spending.

We have attempted to show how a particular government policy, i.e. large-cost ineffective irrigation schemes can worsen the problem of responding to the Dutch-Disease type adjustment. Other agricultural policies have also been implemented in this period affecting supply response to real appreciation of the exchange rate. This brings us to the next major agricultural development strategy of the government which is called Agricultural Development Programmes (ADPs).

The Agricultural Development Programmes

This is a 'minimum-package approach' to rural development which, in the case of Nigeria, was initiated in the early 1970s by the World Bank; it gained momentum throughout the 1980s when marked increases in production of cereals in some of the ADPs was observed. The great advantage of the minimum-package approach though the small farmers, as compared to the large-scale projects of the RBDAs, is its extensive coverage, its adaptability to various farming systems, and greater scope for farmer participation. The approach requires a simultaneous increase in the availability of agricultural inputs, knowledge about farming systems, and access to improved institutional support. The main objectives of the ADPs were an increase in income of the smallholders; an increase of productivity; and an improvement in infrastructure. These objectives were to be achieved through *farm and crop development,* involving support to both rain-fed and *fadama* crops with new improved practices involving extension, greater fertiliser use, new improved seeds, credit, and improved applied research systems; *infrastructural development* through feeder roads, improved rural water supplies, staff headquarters, and improvements in marketing, processing and storage; *institutional support and training,* with trained management, a monitoring and evaluation unit, and establishment of service agencies for the supply of inputs; and *technical assistance,* through the recruitment of consultants for better execution of the project and the training of the project staff. In sum, the project provided for the provision and distribution of inputs, credit and extension, and the construction of feeder roads in the remote rural areas (World Bank 1981a).

By April 1984, the Bank had committed loans totalling $663 million for 12 ADPs with a total project cost of $1.95 billion. Two more ADPs were negotiated with potential loans of $197 million and total costs of $311 million (World Bank 1984b, p. 2). Government commitment to push the development of the ADPs was 2.3 billion naira for the programme under the Fourth Plan (Sano 1983, p. 42) Though only a fraction of this amount was in fact made available due to the financial crisis in 1981 (International Fund for Agricultural Development 1985, p. 67).

It has been reported that in terms of production and income the ADPs have done rather well. In the three pilot projects, production gains of more than 5 per cent per year for crops like maize, millet, and sorghum have been recorded (World Bank 1981b, p. 53). Abalu (1982, p. 126) also reports a compound growth of 5 per cent per year for food crops in the first three projects which are Funtua, Gusau, and Gombe, as compared with about 1 per cent average growth for traditional staple food crops, such as sorghum and millet, for the rest of the country. In terms of growth, the Funtua project has shown 190 per cent growth in maize production during the first three years. At least in the initial years, therefore, the projects appear to have had a good production record.

The main criticism which has been directed to the World Bank ADPs is that, although the target groups included all farming families in the project areas, the resource constraint of the smallholders and institutional difficulties in reaching all the farmers have led to a bias in favour of the larger or so-called 'progressive' farmers. These were farmers who proved to be particularly responsive to the extension advice and introduction of new varieties. It was expected that the emphasis placed on such farmers would result in a demonstration effect, causing smaller farmers to follow in the footsteps of the larger ones (Abalu 1982, pp. 57–60). In 1978–9, almost 20,000 farmers were included in the 'progressive' category in the Funtua project while the rest, the so-called 'traditional' farmers, numbering about 67,000, only received minor attention from the project's extension staff. In the same year, the large-scale and progressive farmers received 61 per cent of all extension visits and were similarly favoured with respect of fertiliser distribution and credit, (D'Silva and Raza 1980, pp. 282–97). In 1970–80 the traditional peasants received on average one-tenth of the quantity of fertiliser distributed to larger farmers and eye-witnesses have observed the greater requirements of the small farmers for fertiliser in Funtua itself (D'Silva and Raza 1980).

Another important criticism of the earlier ADPs has been their advocacy of mono-cropping in a risk-ridden environment in which multiple-cropping acts as an insurance device. In the Sokoto state ADP, the appraisal report has explicitly recognised the error of the earlier mono-cropping recommendation (World Bank 1981a, p. 4). In other words, the World Bank has now recognised the importance of a farming-system approach and project preparations are gradually becoming more flexible. Other criticisms include the problem of a technology-first approach which ignores socioeconomic constraints, hence the inability to see why some farmers are unwilling to take up modern practices; poor research support; the enhanced dependence on foreign management and technology; lack of attention to training Nigerian counterparts; weak credit and co-operative components; and a disproportionate emphasis on fertiliser and extension to the neglect of other 'software' components.

In spite of the above criticisms, the ADPs should be considered a step in the right direction in so far as the smallholders have now become the focus of attention, at least in theory if not quite in practice. The study of the project areas, the experience of working with small farmers, and the realisation of some of the socioeconomic constraints at an official level must be considered as progress. It should be added, however, that the ADP approach to rural development is not the principal strategy of the Nigerian government in spite of some of the promising results achieved so far. The 1982 agricultural budget, for example, allocated only 22 per cent of the total budget for this purpose and the actual allocation was only 17 per cent. The large RBDAs received nearly 50 per cent of total agricultural spending (IFAD 1985, Table 47), even though the ADPs have a much greater coverage both in terms of area and in terms of number of farmers affected.

The Green Revolution Programme

Inaugurated in 1980, this was the third in a series of attempts to increase food production in the 1970s. The earlier efforts were the National Accelerated Food Production Programme (NAFPP) in 1972 and Operation Feed the Nation

(OFN) set up in 1976. OFN was discontinued in 1979 because of poor results. The NAFPP has been continued, however, and in 1980 received 0.5 per cent of federal allocations (Sano 1983, p. 39). The NFAPP put great emphasis on research and extension work on an individual crop basis, while offering a complete package of know-how on techniques, pricing, credit, fertiliser, and distribution systems. The programme was aimed at increasing productivity of both staple foods and import-substituting crops, especially rice and maize, but also sorghum and cassava. The target group, however, was the private large farm, and the intention was to give them more credit through the Credit Guarantee Scheme. By 1983, a total of 380 agro-service centres were established through this programme (Watts 1983, p. 500). It was the NAFPP programme which received further support through the establishment of the National Council for Green Revolution. The Green Revolution Programme also widened the emphasis to cover small farmers.

The actual results of the Green Revolution Programme however, are difficult to assess. The RBDAs are now in charge of operationalising the Green Revolution and the programme has been largely restricted to the areas covered by the RBDAs (Udo 1984). Udo (1985 p. 98) observes:

> At the end of the first full year of the green revolution programme, virtually all the farmers interviewed in the southern states did not benefit from the programme; whereas most of the farmers interviewed in the two northern states benefited considerably from it. One main reason for this apparent disparity in the effectiveness of the green revolution as between the south and the north has to do with the fact that river basin authorities play a major role in operationalising the green revolution programme. And since agricultural development is tied up with the provision of water in that part of the country, the RBDAs have become the main agents for increased food production in that part of the country.

In other words, the Green Revolution Programme has been largely absorbed into the RBDAs, which have already been discussed. The severe problems already indicated for the RBDAs would, therefore, cast doubt on the success of the Green Revolution Programme.

Having discussed the agricultural practices and policies, the necessary background is provided for a discussion of prices, production and income at crop level.

Prices, production and incomes

In this section, relative price, production, and incomes from various crops are discussed in order to achieve an understanding of the process of crop substitution in Nigerian agriculture in the 1970s. The analysis considers both the relative output price of various crops and also, to the extent possible, the input side. The analysis of inputs is largely an analysis of labour requirements at different levels of yield for various crops. The labour-cost data, combined with the price data, will provide a general idea of the incomes obtained from different crops by the farmers. Changes in the relative prices of farm products, therefore, would involve changes in the opportunity cost of labour and the reallocation of labour can explain crop performance accordingly. The expectation from the theoretical framework of Chapter 1 is that all traded crops would decline, although the more the labour requirements of that crop per unit of output, the greater the adverse impact of real appreciation on the value added. With real appreciation and higher prices of non-traded crops, labour will be allocated in that direction. This would bid up the wage rates at full employment The analysis should also take into account perceived risks associated with various crops by farmers, in order better to explain crop performance. Moreover, the previous discussion concerning the government's input and price policies are brought into the discussion at various points to determine the extent to which policy factors affect supply response.

Food crops

There are various estimates for the production of food crops in Nigeria. The three sources of data are the Federal Office of Statistics (FOS), the Food and Agriculture Organisation (FAO), and United States Department of Agriculture (USDA). The FOS uses rural economic survey for its data. The FAO and USDA have their own estimates and did not refer to the FOS figures. They do not even refer to each other for their estimates. Usually the resident mission makes its own estimates by analysing weather reports, satellite imagery, price trends, imports (if present) and field observations. Evidently, there is a data problem (Table B.25).

FOS estimates from 1970 to 1982 show a decline in production of both cereals and root crops. The average annual rate of decline for the seven principal commodities over the 1970–82 period was 1 per cent. This is a weighted average for sorghum, millet, maize, rice, yams, cassava and cocoyams. The decline in production of the three main root crops was a staggering 50 per cent. If this picture is accurate, it would suggest a substantial fall (12.5 per cent at the national level and near 25 per cent in the south) in per-capita calorie intake which would amount to a national crisis and which evidently did not happen in this period.[6] Moreover, the rise in food imports (from $0.2 billion in 1971 to $2.3 billion in 1980), in the same period was insufficient to replace the alleged decline in the production of root crops.[7]

The FAO and USDA statistics, on the other hand, suggest a vastly different picture. Increases in production are recorded for all major food crops since the early 1970s (with the exception of the USDA figure for cassava, which records a reversal of the steady increase since 1970 due to the effects of pests). The FAO and USDA studies usually differ over the estimated production figures although their trends are comparable. There appears to be a regularity in the increases in production reported by both FAO and USDA, indicating that these organisations may have resorted to projections on the basis of a demand-based relationship between growth in production and in population. In any case, their figures are totally inconsistent with those of FOS.

Let us take maize as an example. A new technical package combining fertiliser and improved varieties has been introduced in a number of areas in the north. FOS shows a substantial decline in maize production in 1970s, whereas the FAO and USDA both show rapid increases. The FOS figures for cassava move in the opposite direction to those of FAO. Also, it is not only the direction of change but also the actual volume of production which differ. Again for cassava, the FOS figure for 1970 is only half the FAO figure; the figure is only 15 per cent of the FAO figure. It may be concluded that the data base is weak and contradictory and no conclusions are possible on that basis. This, however, may be a hasty conclusion as recently a consensus has emerged, as reflected in the World Bank (1984a) documents and agreed by the government The consensus appears to be that cereal production increased slightly during the 1970s while root crops showed a slight decline, and the aggregate food production may best be described as stagnant, or having at most an average compound growth of 1 per cent during the decade. By any account, however, the rate of growth in food crops was below the rate of population growth, estimated at 2.5 per cent per year, and per-capita food production did decline (World Bank 1984c).

Two other sources confirm the growth in food crops – P. Clough and P. Collier. Clough in his recent field work in the north of Nigeria has confirmed the growth of millet and sorghum production (Williams 1983). Also Diehl's (1982) study of the southern Guineau savannah reveals a rising trend for yam production due to its high profitability. We now consider a number of specific food crops. The data used are mostly from the USDA since their data comes nearest to the presently agreed state of affairs, though, if more specialised sources are available, the data are complemented accordingly.

Rice. After Sierra Leone and the Ivory Coast, Nigeria is the largest producer of rice in West Africa. According to the USDA, rice production grew from about half a million tons in 1973 to over 1 million tons in 1980 (Table B.15). According to the West Africa Rice Development Association (WARDA 1981, p. 4), paddy output rose by 20 per cent between 1973 and 1978. There is general agreement that rice production showed considerable growth in the 1970s. The main factors in this growth appear to be technological improvements in rice production through improved seed varieties, greater application of fertiliser, and a high demand for rice resulting in incentive prices since rice has a high positive income elasticity of demand. Official rice prices have not risen as fast as the general inflation (Table B.16), but by all accounts the market price of rice has been considerably higher than the official price. Moreover, imports of rice rose from about 5,000 tonnes in 1972 to nearly half a million tonnes in 1978 (WARDA 1981, p. 7). The level of imports, however, was insufficient to maintain a constant domestic price. In fact, domestic prices were on average above the price of imported rice. Between 1973 and 1978, for example, the average price of imported rice (cif) was 350 naira per tonne, while the average wholesale price of rice in the same period in Lagos which is a port city was 577 naira per tonne. The government's rather erratic rice-imports policy has apparently not allowed or not been able to reach a sufficient import volume to equalise the domestic and world price.

There are two basic rice-production systems in Nigeria: upland or rain-fed rice, and lowland or swamp/*fadama* rice. In 1975, upland rice, with a yield of 0.5–0.9 tonnes per hectare for the traditional upland and 1.4–1.5 tonnes per hectare for improved upland, produced about 43 per cent of Nigeria's total rice output (WARDA 1981, p. 9). This system of rice production is predominant in the southern part of Nigeria. Upland rice also exists in the northern part of Nigeria, but its lower yield makes it less popular than *fadama* rice. The NAFPP was important in raising the productivity of some of the upland rice in the south through improved seeds and fertiliser (WARDA 1981, p. 13).

The second most important system of rice production in Nigeria is inland swamp rice along the river valleys in the northern states. Only one crop is planted every year and there is no or little water control. Yields are nearly double those of the upland systems. The recent improvement of *fadama* rice through the ADPs has given yields nearly 50 per cent higher than the traditional yields (WARDA 1981, p. 15). This system covers nearly 30 per cent of the area under rice in Nigeria. Other forms of lowland rice also exist in various parts of Nigeria but are not of significant proportions.

Furthermore, irrigated rice production has recently also developed through the RBDAs. This system has full water control, mechanised land preparation, and makes heavy use of fertiliser. Yields obtained are generally higher than in the other systems and it accounts for about 8 per cent of the area under rice. The high cost of the RBDAs, as discussed previously, casts doubt on the economic viability of such irrigated rice production.

We now consider the cost of rice production. The principal input is labour, which accounts for over 65 per cent of the total cost of all the systems (WARDA 1981, p. 17). Labour requirements of rice production vary – 160 person-days per hectare for traditional upland, 196 person-days per hectare for improved upland, and 118 person-days per hectare for *fadama* in the north, at yield levels already mentioned. Taking farm wages during the 1975–6 production season, which varied from 0.95 naira to 1.15 naira per day, WARDA (1981) found that all rice-production except the upland north system are very profitable.

What emerges from the study of rice is that the rising price, rising yield and, as we shall see, a relatively low labour requirement compared to other food and export crops are the basis for growth in rice.

It is interesting to note the interaction between the income and the price (or substitution) effect in the demand for rice which has become an increasingly traded commodity in Nigeria. On the one hand, the rise in income raises the demand for rice, causing higher rice prices; the higher price is then partly offset by greater levels of imports, greater domestic production, and some substitution away from rice towards cheaper food. Moreover, the appreciation of the exchange rate contributes to greater rice imports by enhancing the income effect through a price effect which involves switching to cheaper imported (and perhaps better-tasting) rice; furthermore, there are cost-reducing technical changes in domestic rice production which increase the domestic output. The net effect on domestic prices and output would, therefore, depend upon the interaction between the income and the price effect influencing domestic prices, and productivity-enhancing technologies.

Maize Maize or corn is the other crop with good production growth. Maize grows in the western part of Nigeria where rainfall permits two crops per year. The weather is the main determinant of yields which are generally low. The expansion of production is the result of both area expansion of production and productivity growth. As mentioned earlier, many of the larger commercial farms produce maize for chicken feed. Also, maize has had a good record in the northern ADPs and this also accounts for its high production growth (Table B.18). Maize is mostly used for livestock and poultry feed and therefore has a high positive income elasticity of demand. Maize imports have been very small relative to domestic production and so maize is mostly a non-traded crop, for this reason we would expect its price to rise in the 1970s. The minimum support prices for maize were raised on average by 9.8 per cent per year during the 1970–81 period. Between 1976 and 1981, however, minimum maize prices were rising faster, at 14.2 per cent per year. Actual purchases by the government, however, were negligible, indicating that the market prices must have been well above the support levels (USDA 1981, p. 13).

Maize provides an example of a non-traded food crop (in the sense that the domestic price is too high to make it an exportable and too low to make it an importable), the actual domestic price of which rises significantly. We have no information concerning the actual market prices of maize: however, it is known that the minimum prices, especially after 1976, were rising along with the rate of inflation and that little was purchased at this price since producers could obtain much higher prices on the open market. Therefore, we can conclude that the market price of maize must have risen well above the rate of inflation, as in the case of other non-traded commodities. Maize production improved not so much because of yield improvement as because of area expansion.

This is another indication that terms of trade for corn must have improved so as to induce the farmers to allocate more of their land for its production.

Also, labour requirements for corn are relatively low – between 110 and 130 person-days per hectare (Diehl 1982, p. 5).

Millet and sorghum. These are the main subsistence crops in the northern parts of Nigeria. In terms of total acreage and per-capita consumption, sorghum is the most important grain in Nigeria (USDA 1981). Most farmers still grow the old indigenous variety which has low yields but grows in impoverished soils and is resistant to major pests. Millet is also a major food crop and it has the advantage of being drought-resistant. Most farmers include millet in their cropping activities and a large part of the millet grown is consumed by the farming household. The rate of growth in both millet and sorghum has been about 1.8 per cent per year and government support prices for both crops rose by more than the average rate of inflation (Tables B.19 and B.20). The food-marketing board involved with millet and sorghum, however, purchases very little of the total production and therefore government support prices are largely irrelevant for the producers; market prices have been well above government prices. Moreover, the labour requirement for millet and sorghum is around 120 person-days per hectare which is again relatively low (NEDECO 1975).

It should be added that subsistence crops such as sorghum and millet are also subject to important income effects since the rise in income induces a shift of demand away from subsistence crops and towards higher income food crops such as wheat, rice, and animal proteins.

Finally, Collier's (1983) exhaustive survey of micro-studies reveals that although agriculture's share in national income declined rapidly in the 1970s, within the agricultural sector, food producers gained relative to export farmers, because of the rise in the relative price of food which remained largely non-traded. It may be misleading, however, to refer to food producers and export producers as two different classes of people since generally farmers combine export crops and food crops within the same farming system. Perhaps a better description of the process is one of crop-substitution within the farming system as opposed to income inequality between food farmers and export farmers. Nevertheless, the point about the rising price of food is a valid and important one which is likely to have resulted in an expansion of land under food crops and rising land prices (where land is bought and sold) in areas which were land short.

Having discussed the production and the relative price movement of some of the main food crops, let us now consider the export crops which, given our analytical framework, would be expected to decline unless they receive crop-level support from the government.

Export crops

We can begin the discussion of the traditional export crops with groundnut, which at present is no longer an export crop in Nigeria but a food crop. Nigeria was the largest groundnut producer in the world and exported an average of 550,000 tonnes per year in the 1960s. From 1970 onwards, however, groundnut exports fell sharply because of low production and high domestic demand, and since 1975 such exports have been banned. Moreover, the world price of groundnut between 1974 and 1980 fell by 34 per cent, thus further discouraging groundnut exports (Table B.21).

Production of groundnut in the 1960s was on average about 1.6 million tonnes per year, while in the 1970s it dropped to under 600,000 tonnes per year. In 1981 and 1982, production was slightly above the 1970's average but well below the 1960's average (Table B.20). It should be noted that the decline in production was due to the decline of both area and yield by about 30 per cent in the 1970s (as compared to the 1960s). Rosette disease in 1975 wiped out the bulk of that year's crop. While groundnut exports were reduced in the 1970s and eliminated in 1975, groundnut production continued, due to demand for it as a food crop destined for domestic processing and consumption. Indeed the combination of high domestic demand and the government's ban on groundnut exports in Nigeria effectively turned groundnuts, a major traded commodity in the 1960's, into a non-traded commodity in the 1970s.

The domestic real price of groundnuts did not decline in the 1970s. From 1970 to 1980, the groundnut producer price rose from 47 naira per tonne to 420 naira per tonne, which is an average growth of 22 per cent year (Table B.24). Given the average inflation of about 15 per cent per year during the 1970s, the groundnut producer price rose by an average of 7 per cent per year. Moreover, these prices are the official ones and they are likely to underestimate the actual market prices.

In spite of the increasingly non-traded character of groundnut, why has production shown a falling trend? It appears that the reason lies in the higher labour demands and the seasonal labour requirements of groundnuts as compared to other crops. The simultaneous planting of groundnuts with other food crops and the extreme importance of timeliness in groundnut cultivation, combined with increasing labour scarcity and rising labour costs relative to other crops, appear to be among the chief reasons behind the decline of groundnut (NEDECO 1975). Moreover, the neglect of export crops, including groundnut, since the 1973 boom, and the preferential distribution of fertiliser to other priority food crops have had a detrimental impact on groundnut production. Under the conditions of severe labour constraints, as discussed earlier, area expansion is no longer feasible but raising land productivity involves modern inputs which for groundnut have not been forthcoming. Higher labour requirements and the absence of policy attention to groundnuts, in spite of their rising price, appear to have worked against the crop. Here is the case of a non-traded commodity which, because of high labour requirements, has shown a real decline.

Palm oil, like groundnuts, has also become a domestic food item. The high domestic demand again resulted in favourable market prices and, as was indicated before, palm oil was the only traditional export crop which showed a slight growth in production (Table B.23).

Palm oil and palm kernels are joint products derived from palm fruit. During most of the twentieth century, palm oil has been Nigeria's largest single export commodity, accounting for 30 per cent of world trade (Kilby 1981, p. 7). The disruption of the palm belt during the war years (1967–9) and unfavourable pricing policies by the government resulted in a major decline in palm-oil exports. Moreover, local production only increased marginally during the 1970s, a period of fast-growing domestic demand for palm oil.

The total area under oil palms is estimated at 2.2 million ha, of which 90 per cent is accounted for by wild palmeries (Kilby 1981). The remaining 10 per cent are large-scale government plantations with improved crop varieties.

About three-quarters of the palm fruit is hand-processed by women with a physical extraction efficiency of 60 per cent (against a potential of 97 per cent in modern mills). In other words, Nigeria is losing a vast amount of palm oil each year due to physically inefficient extraction methods.[9]

Since 1960, marketing board purchases have continuously declined due to low prices offered to harvesters. The price of palm fruit in 1972 was set at 40 naira per tonne of fruit in season and 45 naira out of season. These prices remained unchanged in 1980. Between 1972 and 1980 the domestic price of palm oil rose some 295 per cent and kernel palm prices some 195 per cent (Table B.3). During the same period CPI rose some 247 per cent. Due to these pricing policies, the government-operated pioneer mills have been able to attract only enough fruit to operate at 5-10 per cent of single-shift capacity (Kilby 1981, p. 10). Private mills, on the other hand, have fared much better. In other words, the official producer price has not been effective and the domestic price has been much higher.

Moreover, internal wholesale prices for palm oil in 1977–9 were on average two or three times the import parity price (Table B.24). In fact, the world price of palm oil between 1974 and 1989 fell nominally by about 13 per cent (Table B.21). The oil-palm sector and palm milling have been protected mainly through ever-growing domestic demand (3.5 per cent per year) outstripping a fairly stagnant domestic supply (Kilby 1981). The real domestic price of palm oil rose sharply from 1972 to 1976. Since then, however, it has been declining continuously due to ever-increasing imports. More recently, edible-oil imports have been put on an open general licence, which implies an effective end to restrictions on oil imports. Concerning palm kernels, the Palm Produce Board in recent years has purchased palm kernels from producers at prices well above the export parity price.[10]

Moreover, the exports of both palm kernels and palm-kernel oil, neither of which are in much demand inside Nigeria, are subsidised. The price offered by the Board to producers in 1981 was 200 naira per tonne while palm kernels were selling in Europe at a price equivalent to 170 naira per tonne. Including the handling and transport cost, the Board's total cost per tonne of palm-oil kernel exported was 250 naira per ton, giving an export subsidy of 47 per cent of the price CIF Europe or 67 per cent of the export parity price of 120 naira per tonne. In spite of this subsidy, many palm-oil producers, rather than sell to the Palm Produce Board, find it more profitable to use the kernels to fuel their boilers.

The subsidy to palm kernels also involves a subsidy to palm-oil production since they are joint products. Given the presence of import controls, one can see how the palm economy has benefited on two fronts. Import controls on vegetable oils, at least until recently, had turned palm oil into a non-traded commodity. Higher domestic income and demand raised the price of palm oil far above the import parity price level while subsidisation added a further incentive to palm-fruit milling in particular and the palm-produce economy in general.

As was mentioned earlier, palm-oil production appears to have shown only marginal growth. Moreover, there has been a shift of production away from large and medium-scale commercial mills to small-scale artisanal operations.

The large establishments have been unable to prosper because of the shortages of raw materials and/or pricing policy constraints which have prevented them from operating at full capacity.

In sum, the income effect and the increased expenditure financed by oil have resulted in the rapid rise of domestic demand for this commodity. Import controls resulted in palm oil becoming a non-traded commodity, thus its price has risen substantially above the international price. In spite of the favourable price, palm-oil production has not increased since the large and modern crushers with high extractive capacity have been unable to secure a sufficient supply of palm fruit due to poor pricing policy. In turn, the more traditional crushers have received the bulk of palm fruits but they can only extract at a much lower level – technical improvement has not taken place as far as palm oil is concerned.

The analysis of palm oil shows the fundamental importance of distinguishing between the price and the income effect and of being careful not to apply the Dutch-Disease hypothesis in its free-trade form. Douglas Rimmer (1984, p. 244), for example, has incorrectly attributed the decline of palm-oil exports to the price effect of a higher exchange rate and the Dutch-Disease phenomenon in its free-trade form. He writes that palm produce and groundnut products exports have declined since the expenditure of oil revenues 'raised the opportunity cost of labour and also led, through inflation, to the overvaluation of the naira'. A recent agricultural atlas of Nigeria by Agbola (1979) makes a similar incorrect assessment since it argues (1979, p. 121) that 'the export of palm oil has declined as a result of increasing production cost during a period of falling producer prices'.

As the analysis has shown, however, both Rimmer and Agbola have an incorrect assessment of the reasons behind the decline of palm oil exports.

Rimmer has assumed that palm oil is a traded commodity and that, due to the real appreciation of the naira or higher domestic cost relative to price, it has become unprofitable to produce palm oil. He also takes the decline of exports as indicating the decline of production in general. In fact, as we have seen, due to import controls, palm oil has been a non-traded commodity and its domestic price has been quite high. Moreover, palm-oil production has not fallen but has remained stagnant and this has to do partly with poor pricing policy by the government which has led to the use of traditional extraction methods with lower extraction rates.

In addition, higher prices cannot substantially increase supply unless there is a large-scale attempt to replant the present wild palmeries with high-yielding varieties. Furthermore, as we have seen exports have fallen because of the rechannelling of palm oil into the domestic market which was until recently sheltered from vegetable-oil imports. Finally, if one applies the Dutch-Disease idea, as Rimmer does, or argues that price/cost ratios have fallen, as Agbola does, one should also expect a fall in palm-kernel exports. Palm-kernel exports increased substantially between 1974 and 1976 thanks to the government export-price subsidy and their level throughout the 1970s was continuously above their 1970 level (Table B.24). Both palm oil and palm kernels show the critical role of the government policy *vis-à-vis* these crops since palm oil is non-traded due to government import control and palm kernel is subsidised.

Another important point concerning palm oil is its high labour requirement both for harvesting and for crushing. Mitra (1972) found that palm-oil harvesting and processing requires 40 per cent more person-days per hectare than yams and 80 per cent more than cassava. Given the inter-cropping of palm oil with yams, it is not difficult to see the shift of labour away from oil towards the rootcrops in the south.[11]

Next let us discuss cocoa. Among the traditional export crops, cocoa beans are now the only significant non-oil foreign exchange earners. In 1979, cocoa beans accounted for more than 60 per cent of the total value of non-oil exports. The increase in the foreign exchange earnings of cocoa has largely depended on price developments abroad rather than volume changes of Nigerian exports.

Cocoa production has declined significantly. From an average of 240,000 tonnes between 1968 and 1973, it fell to an average of 170,000 tonnes between 1974 and 1980 (Table B.27). Area and yield showed a steady decline throughout the 1970s. Looking at the real price of cocoa, we find that although it rose in 1974, the price rise was gradually eroded by the rising inflation of the 1970s. Moreover, the price rise in 1977 did not even restore the real price of cocoa to its 1973 level and this price rise was again eroded by inflation (Table B.27). In other words, although the price of cocoa twice rose sharply, in 1974 and in 1977, the real price was consistently below the 1974 level throughout the post-1974 period. This reflected the process of real appreciation which involves the rising relative price of non-traded goods. We may conclude,

151

more, the price rises were partly due to favourable price developments abroad which allowed the marketing board to raise the domestic price. What is important for our purposes, however, it not the sudden rises in the price but the trend of the real price which was indeed downwards in the 1970s.

Exports fell in the 1974–80, as compared with the 1968–73 period. Although the total value of cocoa exports more than doubled between 1974 and 1980, the actual quantity exported declined by about 10 per cent. Moreover, cocoa has an extremely high labour requirement in terms of person days per hectare, almost five times that of cassava (Mitra 1972). This means that, given the falling real price, return to labour in cocoa production has fallen to very low levels in the 1970s, hence the drop in production.

Summary

The rising profitability of food crops resulted in a process of crop-substitution in Nigerian agriculture in the 1970s. Food crops, with the exception of wheat and rice, appeared to be non-traded at the margin, hence their prices rose in line with or even above the rate of inflation. Moreover, food crops such as millet and sorghum in the north have lower labour requirements than groundnuts. Also the simultaneous planting time of groundnuts with other food crops in the north, and the relatively high labour requirement of the former, resulted in a shift of labour away from groundnuts and towards millet and sorghum. Moreover, the lower labour requirement of root crops in the south and their high price, resulted in a shift of labour away from palm-oil and cocoa production towards cassava and yams (both non-traded crops). It should be added that the low growth rate of both cassava and yams is attributable to the absence of an easily accessible productivity enhancing input package for the Nigerian root-crop producer. It should also be added that rising land prices in this period increased the cost of production with adverse effects on incentives for expending the level of output, thus cancelling part of the improved profitability due to the higher price of non-traded food crops.

Putting the above crops and their price levels into the traded/non-traded diagram of Chapter 1 (Figure 1.1), the horizontal spectrum shown in Figure 6.2 emerges for the pre-1973 and post-1973 periods in Nigerian agriculture (see also Figure 4.1 in Chapter 4 on Iran).

Notes

1. The World Bank (1984a, p. 86) has estimated that there are 9 million families and 2.5 ha is the average size. According to the rural economic survey in 1980, there were 6.9 million farming families while the average farm area for rural Nigeria was 0.5, 0.6, and 0.9 ha in 1977, 1978 and 1979, respectively (see Federal Office of Statistics (FOS 1977–8, 1978–9, and 1980–1.
2. An interesting difference concerning the studies of the agricultural sector of Iran and

Nigeria which the author has noted is the relative abundance of literature on farming systems in Nigeria while it has been difficult to find any literature concerning farming systems in Iran. It appears that the food-crop export-crop dichotomy and its management at the farm level is far more significant for the Nigerian farmer than for the Iranian farmer. This may be the result of the large-scale imposition by the British of an export-crop economy which resulted in the creation of such a sharp dichotomy. In Iran, although the food-crop–cash-crop distinction is important, it usually does not refer to two different crops but the same crop produced at different levels of output.

3. The description of the institutional set-up of the RBDAs is based on World Bank (1979, Vol. 3, Paper 6).
4. For the Bakolori project, for example, the FAO recommended a pilot project, but this was never undertaken, see World Bank (1979).
5. Assuming the whole area is devoted to rice and assuming that the economic value of rice is $750 per tonne, the yield of rice has to be 24 tonnes per hectare to reach the break-even point. In 1981, the average yield of rice for the RBDAs as a whole was around 2 tonnes per hectare.
6. The proportion of calories supplied by root crops is from FAO/ESP, Country Tables, 1983.
7. The view on insufficiency of food imports to replace for the alleged fall in domestic food production is adopted from the World Bank (1984d, p. 7).
8. Wheat has not been discussed in any detail since Nigeria produces an insignificant amount for its own consumption. Moreover, most observers agree that agro-ecological conditions in most parts of Nigeria are inappropriate for wheat production. As was mentioned earlier in the discussion of the RBDAs, wheat is produced at high cost with heavy subsidy.
9. Whether or not such traditional methods are also economically inefficient depends on the cost of modern mills compared to the traditional mill and the shadow pricing of labour and capital for which we do not have sufficient data.
10. The following section on Palm Kernels is based on World Bank (1981c).
11. Lagemann (1977, p. 228) has shown that in this period yams became the most profitable crop.

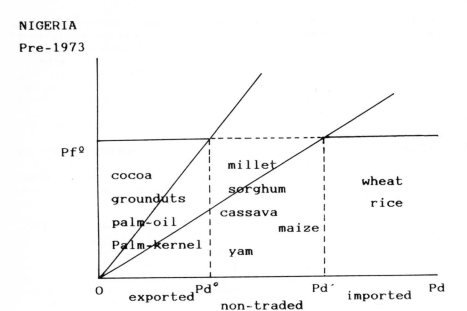
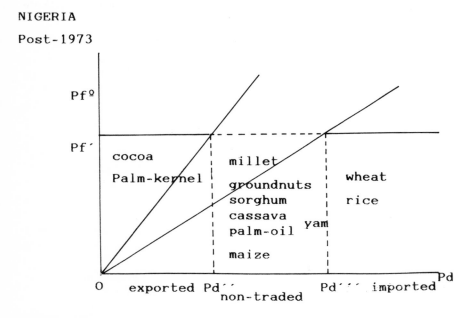

Figure 6.2

7 Conclusions

This book has discussed the theoretical structure and two empirical applications of the so-called Dutch-Disease model, which has been shown to provide a convenient initial framework for the analysis of macroeconomic and sectoral price and output developments in the oil-based economies of Iran and Nigeria during the 1973-80 period. Both countries followed a high-expenditure path in order to adjust to the accumulation of financial surpluses in their balance of payments. The high levels of expenditure in both cases were the result of the rapid increase in public and capital expenditure, especially in the non-traded sectors such as infrastructure, housing and transport. The increase in public expenditure was then followed by an increase in private expenditure. The rise in the overall aggregate expenditure meant an increase in the relative price of non-traded goods.

The Dutch-Disease model, which is based upon a three-factor two-commodity model of production and trade, predicts the following inequality

$$R_n\sim > P_n\sim > W\sim > P_t\sim > R_t\sim$$

where $\sim$ refers to relative change, Ri and Pi refer to the rental on capital and the relative price in the non-traded and traded sectors ($i_i=n_i$, t), and W refers to the wage rate which is equalised across the economy.

The observed outcomes, however, were somewhat different, and we have argued that these differences arise mainly because of government intervention. In Iran the observed inequality was:

$$P_c\sim\, >\, W\sim\, >\, P_s\sim\, =\, P_a\sim\, >\, P_m\sim$$

where $P_i\sim$ denoted the percentage change of the GDP deflator for the construction, services, manufacturing and agriculture sectors (i = c, s, m, a), and $W\sim$ refers to the percentage change in average nominal wage rates in the different sectors of the economy. In other words, in Iran the real wage rose relative to agriculture, manufacturing and services but fell relative to construction. The rise in the real wage relative to agriculture reflects the cheap-food policy in Iran which was made possible by the financial strength of the government and the largely traded character of food in Iran (especially wheat). Also, services were very much influenced by direct government intervention in the form of consumer price studies on public transport and the largely share of government in the total value added in the service sector. In other words, in spite of the non-traded character of services, their prices did not rise above the wage rates due to government intervention.

In Nigeria, the inequality took a different form:

$$P_a\sim\, =\, P_s\sim\, =\, W\sim\, >\, P_m\sim\, >\, P_c\sim$$

The rapid price rise in the agricultural sector can be explained largely in terms of the non-traded character of food staples in Nigeria and the absence of an effective cheap-food policy. Wages rose practically parallel to the rise in food prices and so did the price of services, reflecting the fact that, in large part, income from services is identical to wages because of the self-employed and non-traded character of most services. It should be added that, although government services form about a third of the value added in services, the rapid rise in the public sector's salaries and wages further raised relative prices in this sector and the government did not implement any effective subsidy policy directly focused on the service sector.

On the output side, the role of the government intervention was shown to be critical since, with the exception of construction in Iran, the sectors with the lowest price rises showed the fastest real growth rates. In Nigeria, in contrast to Iran, the bulk of construction activities were non-residential buildings, and specifically in publicly sponsored construction in physical infrastructure. Moreover, protection and promotion of selected manufacturing activities allowed maintenance of high profit margins caused by the faster growth of prices than of wages. Moreover, it was shown that the presence of traded inputs which substitute for non-traded inputs can partly offset the adverse impact of real appreciation on the traded sector and in this sense the role of capital inputs (such as land and imported machinery) was also introduced into the analysis.

The principal conclusion of this book is that the Dutch-Disease model can be usefully complemented, and its usefulness enhanced if the role of government and capital inputs is introduced into the analysis for explaining output performance in oil-based economies. Moreover, if the Dutch-Disease model of 'oil economy' is used without the introduction of these complementary factors for an explanation of the consequences of oil boom and oil doom, the result can be misleading precisely because alternative policy choices which could have altered the outcome would not have been adequately specified for the economies in question. It is not uncommon to find analysis of oil economies which at most limit the policy discussion to specifying the macro-level fiscal, monetary, and exchange-rate policy, while sectoral policies, investment projects, and the form of government intervention are not explicitly considered and the role of capital inputs not adequately analysed.

A supplementary finding of the book is that the question of what is traded and what is non-traded cannot be determined a priori. The usual classification of agriculture and manufacturing as traded and of construction and services as non-traded seems to be inadequate. We have seen how within agriculture there are important non-traded crops, such as key staple foods in Nigeria; also within industry, or rather manufacturing, there are a number of construction related products with high bulk and relatively low value, such as bricks, which are also largely non-traded. Also heavy product differentiation, for example Persian carpets, can make a product exportable but not importable, which is a rather unusual combination. Moreover, commodities move from being exported to being non-traded and on to being imported as their domestic price levels rise due to the expansion of domestic demand, and therefore static categorisation of commodities or sectors can be misleading.

Another important result is that the relative price adjustment and the resource shift towards non-traded goods is not a smooth process and appears highly erratic. Growth rates of prices and output differ greatly year by year in all of the sectors of the economy (specially in the construction sector). The fluctuation is due both to government policy and to changes in oil revenues.

Furthermore, it is important carefully to distinguish the decline of exports due to appreciation of the exchange rate or the falling cost/price ratio and the decline in exports due to the expansion of domestic demand. A number of commentators seem to emphasise the price effect caused by higher real exchange rates whereas in fact it has been the income effect and the expansion of the domestic market which is the reason for the decline in exports.

Before ending this discussion, an important underlying theoretical point requires further elaboration. This point is related to the placing of the oil-exporting countries within a long-term generalisation about the development process.

Kravis *et al* (1981) have attempted to analyse the long-term structural factors which determine the relative price level of a country. Thirty-four International Comparison Project (ICP) countries were arranged in order of increasing real GDP per capita. Price levels were calculated by dividing the GDP or any of its components at international prices. The results showed that overall price level rises systematically with GDP per capita across countries. Moreover, it was shown that the difference between the price level of tradables and non-tradables clearly increases as income rises.

One explanation for this systematic result goes back at least to Ricardo.

> "The phenomenon can be explained in terms of the difference in the productivity gap between high and low income countries for tradable and non-tradable goods. International trade tends to drive the price of tradable goods, mainly commodities, towards equality in different countries. With equal or nearly equal prices, wages in tradable good industries in each country will depend upon productivity. Wages established in the tradable good industries within each country will prevail in the country's non-tradable goods industries. In non-tradable goods industries, however, international productivity differentials tend to be smaller. Consequently, in a high productivity country high wages lead to high prices of services and other non-tradable goods, whereas in a low productivity country low wages give rise to low prices of services and other non-tradables. The lower the country's income, the lower will be the prices of non-tradable goods and the greater will be the tendency for the exchange-rate conversions to underestimate real income compared with that of richer countries (Kravis *et al.* 1981).

In this book we have seen how in an oil exporting country the prices of non-traded goods rise but this time not due to productivity differentials between the two sectors (supply factors) but mainly by demand factors (supply inelastic of non-traded goods plus income effects). That largely demand factors push the relative price structure of oil-exporting countries towards the pattern of the more developed countries implies that the traded goods sector will undergo extreme pressure from the cost side. If the required technical change and/or direct government action is not forthcoming, traded goods are bound to decline. We have referred to the difficulty of trying to overcome rising costs by price protection and the costs that this involves for the consumer. This leaves us with either lower levels of domestic absorption of oil revenues or the required technical change without which the traded goods producing sector will decline.

Finally, the principal policy conclusions which emerge from this book are the following:

1 The real appreciation consequent to adjusting to higher foreign exchange earnings should be accompanied by measures on the supply side so as to assist those traded industries or agriculture which can have comparative advantage with lower levels of foreign exchange earnings; the cost of this assistance may be justified in terms of an insurance premium for the country as a whole;

2 The level of expenditure chosen should consider the limits to efficient investment set, among other reasons, by the rapidly import-cost problems which arise due to the absence of domestic substitutes and the rise in labour costs. Real appreciation should at best take place slowly since the real appreciation of the exchange rate is required for adjusting to the balance-of-payments surplus. The pace of increased expenditure can be controlled, for

example, by greater levels of investment through investments abroad or lower levels of depletion.
3. The composition of expenditure affects the supply side of the economy significantly and the choice of where to spend must be studied so as to also benefit those sectors with non-oil comparative advantage which may be squeezed in the adjustment process, namely the traditional traded goods producing sector.
4. Large projects and high levels of expenditure go hand in hand, and the choice of type of project has both sectoral as well as macro impact. A small-farmer emphasis in the agricultural sector, instead of an emphasis on large-scale government-managed projects, could in principle result in lower outlays and better sustainability and better impact on income distribution and food production. This approach would also improve the management constraint which is a highly scarce good in the developing economies. In sum, the principal policy aim should be how to develop a viable and diversified economy by using oil in order to grow out of it.

STATISTICAL APPENDIX A: IRAN

STATISTICAL APPENDIX A

Table A.1 Oil exports (millions of US dollars)

Year	Total exports	Export of oil and gas	% of oil and gas in total exports	Production of crude oil (billion barrels)
1969–70	1,519	1,099	72	
1970–71	1,690	1,274	75	
1971–72	2,734	2,150	78	
1972–73	3,337	2,597	77	
1973–74	6,232	5,160	82	
1974–75	20,922	19,235	91	2.2
1975–76	21,971	19,255	87	1.9
1976–77	24,618	20,854	84	2.2
1977–78	25,590	21,096	82	2.0
1978–79	22,738	18,116	79	1.6
1979–80	22,658	19,386	85	1.3
1980–81	14,214	11,851	83	0.5
1981–82	14,320	12,456	86	0.5
1982–83	21,455	20,050	93	1.0

Source: Bank Marzaki Iran, *Annual Reports* 1970/1–1983/4.

Table A.2 Government revenues (billions of rials, current prices)

	Total	Oil and Gas	(%)	Others	(%)
1970–71	172.3	84.7	49.2	87.6	50.8
1971–72	258.9	155.3	60.0	103.6	40.0
1972–73	302.1	178.2	59.0	123.9	41.0
1973–74	464.8	311.2	67.0	153.6	33.0
1974–75	1,394.4	1205.2	86.4	189.2	13.6
1975–76	1,582.1	1246.8	78.8	335.3	21.2
1976–77	1,836.4	1421.5	77.4	414.9	22.6
1977–78	2.034.2	1497.8	73.6	536.4	26.4
1978–79	1,598.9	1013.2	63.4	585.7	36.6
1979–80	1,699.9	1219.7	71.8	480.2	28.2
1980–81	1,325.9	888.8	67.0	437.1	33.0
1981–82	1,651.9	937.9	56.0	713.7	44.0
1982–83	2,375.9	1563.5	65.8	812.4	34.2

Source: Bank Marzaki Iran, *Annual Report and Balance Sheet* 1973–4; 1977–78; 1980–1.

Table A.3 Government domestic deficit (billions of rials)

	1962–63	1963–64	1964–65	1965–66	1966–67	1967–68	1968–69	1969–70	1970–71	1971–72
Total revenues	53.0	59.2	69.1	90.5	97.4	106.5	126.6	152.2	172.3	258.9
Revenues from abroad (mainly oil and gas)	23.7	27.6	36.4	50.0	48.6	54.1	61.8	76.4	84.7	155.3
Domestic revenues (mainly taxes)	29.3	31.6	32.7	40.5	48.8	52.4	64.8	75.8	87.6	103.6
Total expenditure	57.9	63.8	75.5	100.8	107.7	137.0	183.5	205.8	231.8	291.8
Payments abroad	11.9	14.0	18.3	28.1	29.1	43.4	66.2	84.5	105.9	147.1
Domestic expenditure	46.0	49.8	57.2	72.7	78.6	93.6	117.3	121.3	125.9	144.7
Domestic deficit (surplus)	−16.7	−18.2	−24.5	−32.2	−29.8	−41.2	−52.5	−45.5	−38.3	−41.1

Table A.3 Cont'd

(Billion rials)

	1972–73	1973–74	1974–75	1975–76	1976–77	1977–78	1978–79	1979–80	1980–81
1–Total revenues (2+3)	302.1	464.8	1,394.4	1,582.1	1,836.4	2,034.2	1,598.9	1,699.9	1,325.9
2–Revenues from abroad (mainly oil & gas)	178.2	311.3	1205.2	1246.8	1421.5	1497.8	1013.2	1219.7	888.8
3–Domestic revenues (mainly taxes)	123.9	153.5	186.6	324.3	396.7	516.0	564.5	454.6	437.1
4–Total expenditure (5+6)	359.1	478.0	1,254.4	1,775.9	2,006.2	2,492.2	2,207.8	2,227.9	2,298.4
5–Payments abroad	167.0	225.9	696.7	946.9	884.3	969.8	697.8	464.4	498.2
6–Domestic expenditure	192.1	252.1	557.7	829.0	1,121.9	1,522.4	1,510.0	1,763.5	1,800.2
7–Domestic deficit or surplus (3-6)	−68.2	−98.6	−371.1	−504.7	−725.2	−1,006.4	−945.5	−1,308.9	−1,363.1

Source: Calculated from Bank Marzaki Iran, *Annual Report and Balance Sheet*, 1964–5; 1968–9; 1972–3; 1977–8; 1980–1.

Table A.4 Non-government factors affecting changes in the money supply (billions of rials)

	Bank credit to the private sector	Private sector's balance of payments	Net liquidity contribution of the private sector
1962–63	61.5	−18.6	42.9
1963–64	72.9	−22.1	50.8
1964–65	88.1	−35.0	53.1
1965–66	102.2	−33.2	69.0
1966–67	121.2	−39.6	81.6
1967–68	142.3	−40.7	101.6
1968–69	167.2	−50.0	117.2
1969–70	198.6	−56.7	141.9
1970–71	230.2	−64.2	166.0
1971–72	277.6	−67.4	210.2
1972–73	364.7	−82.8	281.9
1973–74	494.2	−149.8	344.4
1974–75	704.6	−328.0	376.4
1975–76	1092.5	−574.8	517.7
1976–77	1516.6	−671.5	845.1
1977–78	1868.8	−687.5	1,181.3
1978–79	2199.0	−645.3	1,553.7
1979–80	2577.4	−701.3	1,876.1
1980–81	3060.4	−825.8	2,234.6

Source: Bank Marzaki Iran, *Annual Report and Balance Sheet* 1964–5, 1968–9, 1972–3, 1977–8, 1980–1.

Table A.5 Components of the money supply (billions of rials)

	Currency plus demand deposits (M_1)	Time and saving deposits	Money supply (M_2)
1962–63	43.9	24.8	68.7
1963–64	49.0	32.5	81.5
1964–65	53.9	38.4	92.3
1965–66	60.3	45.2	105.5
1966–67	66.8	53.9	120.7
1967–68	77.1	67.2	144.3
1968–69	87.9	87.4	175.3
1969–70	90.4	115.3	205.7
1970–71	97.4	138.3	235.7
1971–72	117.0	179.3	296.3
1972–73	158.7	240.7	399.4
1973–74	202.7	313.1	515.8
1974–75	327.2	482.9	810.1
1975–76	446.5	699.0	1,145.5
1976–77	611.2	982.3	1,593.5
1977–78	790.5	1,306.5	2,097.0
1978–79	1,236.5	1,342.1	2,578.6
1979–80	1,665.8	1,884.2	3,550.0
1980–81	2,203.3	2,134.5	4,337.8

Source: Bank Marzaki Iran, *Annual Report and Balance Sheet* 1964–5, 1968–9, 1972–3, 1977–8, 1980–1.

Table A.6 Consumption expenditures and capital formation (billions of rials)

	Private consumption expenditures		Public consumption expenditures		Gross domestic fixed capital formation	
	Current	Constant	Current	Constant	Current	Constant
1970–71	537.3	774.2	141.6	216.0	167.3	261.9
1971–72	566.9	773.4	189.4	277.1	216.7	331.8
1972–73	686.6	880.3	252.6	354.2	287.4	410.5
1973–74	879.7	1,014.6	325.4	427.9	363.3	456.6
1974–75	1,127.8	1,127.8	628.3	628.3	562.0	562.0
1975–76	1,316.0	1,207.3	807.4	722.5	1,056.6	923.6
1976–77	1,869.3	1,554.6	1,003.7	796.0	1,588.8	1,181.2
1977–78	2,487.5	1,684.0	1,133.7	836.9	1,784.6	1,083.1
1978–79	2,932.5	1,798.0	1,255.0	856.4	1,556.8	858.3
1979–80	2,492.3	1,530.6	1,223.1	711.5	1,252.8	623.2
1980–81	3,369.6	1,383.2	1,381.2	770.5	1,476.2	608.8
Compound growth rate (%)	18.2	5.4	23.0	12.3	21.9	8.0

Source: Bank Markazi Iran, *Annual Report and Balance Sheet* 1973–4, 1977–8, 1980–1.

Table A.7 Price level, food prices and real appreciation exchange rate

	IMF–Iran Consumer price index	Food–Price CPI Price Relative	WLD Prices US Index	Real appreciation exchange rate*
1968–69	77.9	113.5	78.2	90.9
1969–70	80.7	99.3	82.4	88.6
1970–71	82.0	98.3	87.3	85.1
1971–72	85.5	100.6	91.0	85.1
1972–73	91.1	102.6	94.1	88.3
1973–74	100.0	100.0	100.0	100.0
1974–75	114.3	101.3	110.9	105.1
1975–76	128.8	100.9	121.1	107.8
1976–77	143.4	96.8	128.1	109.6
1977–78	182.6	90.4	136.4	130.9
1978–79	203.8	96.2	146.7	135.3
1979–80	225.2	106.6	163.3	134.9
1980–81	271.9	113.5	185.4	142.7
1981–82	337.6	118.5	204.6	144.6

Sources: Iran CPI from IMF 1984; food-price index calculated from ILO *Yearbook*; and WLD index from World Bank Country Tables.
* The real appreciation of the exchange rate is calculated as $\frac{\text{consumer price index}}{\text{US price index}} \times$ exchange rate (unit/$).

Table A.8 Gross Domestic Product (billions of rials)

	Total GDP (incl. oil)			Total GDP (excl. oil)		
	Current	Constant 1974 prices	GDP deflator (%)	Current	Constant	GDP deflator (%)
1962–63	331.0	861.2		284.9		495.3
1963–64	350.3 (5.8)	912.1 (5.9)	–0.1	300.7 (5.5)	521.5 (5.3)	0.2
1964–65	396.1 (13.1)	989.4 (8.5)	4.6	339.4 (12.9)	559.9 (7.4)	5.5
1965–66	449.7 (13.5)	1146.3 (15.9)	–2.3	384.4 (13.3)	632.1 (12.9)	0.4
1966–67	487.1 (8.3)	1259.8 (9.9)	–1.6	412.5 (7.3)	677.0 (7.1)	0.2
1967–68	538.0 (10.4)	1407.7 (11.7)	–1.3	451.7 (9.5)	748.9 (10.6)	-1.1
1968–69	613.2 (14.0)	1594.9 (13.3)	0.7	513.1 (13.6)	830.8 (10.9)	2.7
1969–70	690.7 (12.6)	1817.6 (14.0)	–1.3	571.3 (11.3)	899.1 (8.2)	3.1
1970–71	784.1 (13.5)	2045.5 (12.5)	1.0	643.4 (12.6)	987.6 (9.8)	2.8
1971–72	948.0 (21.0)	2345.5 (14.7)	6.4	736.3 (14.4)	1077.3 (9.1)	5.4
1972–73	1190.2 (25.4)	2567.1 (9.5)	15.9	926.2 (25.8)	1233.8 (14.5)	11.3
1973–74	1783.6 (49.9)	2874.4 (12.0)	37.9	1196.1 (29.1)	1423.8 (15.4)	13.7
1974–75	3071.9 (72.2)	3071.9 (6.9)	65.4	1630.3 (36.3)	1630.3 (14.5)	21.8

Table A.8 Cont'd

(billions of Rials)

	Total GDP (incl. oil)			Total GDP (excl. oil)		
	Current	Constant	GDP deflator(%)	Current	Constant	GDP deflator(%)
1975–76	3479.0 (13.3)	3150.1 (2.5)	10.7	2103.2 (29.0)	1885.6 (15.7)	13.3
1976–77	4480.3 (28.8)	3529.9 (12.1)	16.7	2802.2 (33.2)	2145.3 (13.8)	19.5
1977–78	5396.1 (20.4)	3742.6 (6.0)	14.4	3641.1 (29.9)	2379.9 (10.9)	19.0
1978–79	4917.4 (–8.9)	3198.5 (–14.5)	5.7	3691.9 (1.4)	2208.7 (–7.2)	8.6
1979–80	6053.3 (23.1)	3069.5 (–4.0)	27.1	4377.3 (18.6)	2315.3 (4.8)	13.7
1980–81	6148.8 (1.6)	2563.4 (–16.5)	18.1	5312.0 (21.4)	2356.3 (1.8)	19.6

Source: As Table A.5.
Note: 1. Figures in parentheses refer to percentage changes.
2. The GDP deflator has been derived by subtracting the percentage change in current prices from the percentage change in constant prices.
3. Real GDP based on 1974 prices.

Table A.9 Components of GDP (1962/63 –1980/81) (billion rials)

	Agriculture			Manufacturing and mines (excl. oil)			Construction			Services		
	Current	Constant 1974 prices	GDP deflator(%)	Current	Constant 1974 prices	GDP deflator(%)	Current	Constant 1974 prices	GDP deflator(%)	Current	Constant 1974 prices	GDP deflator (%)
1962-63	96.9	185.3		41.8	66.6		13.8	40.2		129.9	201.3	
1963-64	98.4	186.9	0.6	45.4	72.6	-0.4	15.6	46.6	-2.9	137.0	211.9	0.2
	(1.5)	(0.9)		(8.6)	(9.0)		(13.0)	(15.9)		(5.5)	(5.3)	
1964-65	110.6	191.3	10.0	50.3	76.0	6.1	17.6	48.9	7.9	156.0	239.7	0.8
	(12.4)	(2.4)		(10.8)	(4.7)		(12.8)	(4.9)		(13.9)	(13.1)	
1965-66	120.0	205.3	1.2	57.6	86.5	0.7	23.5	63.3	4.1	178.3	272.4	0.7
	(8.5)	(7.3)		(14.5)	(13.8)		(33.5)	(29.4)		(14.3)	(13.6)	
1966-67	121.7	211.8	-1.8	66.4	101.2	-1.7	23.0	61.0	1.5	195.3	396.9	-36.2
	(1.4)	(3.2)		(15.3)	(17.0)		(-2.1)	(-3.6)		(9.5)	(45.7)	
1967-68	128.4	228.2	-2.2	76.0	116.4	-0.5	28.4	71.2	6.8	211.5	325.7	26.2
	(5.5)	(7.7)		(14.5)	(15.0)		(23.5)	(16.7)		(8.3)	(-17.9)	
1968-69	139.6	246.1	0.9	88.2	133.7	1.2	33.1	74.7	11.6	243.2	367.5	2.2
	(8.7)	(7.8)		(16.1)	(14.9)		(16.5)	(4.9)		(15.0)	(12.8)	
1969-70	147.8	253.3	3.0	100.5	147.0	4.0	38.6	75.8	15.1	273.2	412.1	0.2
	(5.9)	(2.9)		(13.9)	(9.9)		(16.6)	(1.5)		(12.3)	(12.1)	
1970-71	160.6	264.7	4.2	113.7	163.1	2.1	41.0	78.1	3.2	314.7	468.6	1.5
	(8.7)	(4.5)		(13.1)	(11.0)		(6.2)	(3.0)		(15.2)	(13.7)	
1971-72	172.7	256.7	10.4	138.1	190.5	4.7	45.1	83.5	3.1	364.5	529.7	2.8
	(7.5)	(-2.9)		(21.5)	(16.8)		(10.0)	(6.9)		(15.8)	(13.0)	
1972-73	201.8	271.0	11.4	171.5	224.8	6.2	58.4	91.4	20.0	477.2	629.4	12.1
	(16.9)	(5.5)		(24.2)	(18.0)		(29.5)	(9.5)		(30.9)	(18.8)	
1973-74	234.4	286.5	10.5	231.9	264.4	17.6	78.9	101.7	23.8	629.3	749.6	12.9
	(16.2)	(5.7)		(35.2)	(17.6)		(35.1)	(11.3)		(31.9)	(19.0)	
1974-75	303.3	303.3	23.5	312.9	312.9	16.6	98.2	98.2	27.9	889.1	889.1	22.7
	(29.4)	(5.9)		(34.9)	(18.3)		(24.5)	(-3.4)		(41.3)	(18.6)	

Table A.9 Contd

(billion of rials)

	Agriculture			Manufacturing and mines (excl. oil)			Construction			Services		
	Current	Constant 1974 prices	GDP deflator(%)	Current	Constant 1974 prices	GDP deflator(%)	Current	Constant 1974 prices	GDP deflator(%)	Current	Constant 1974 prices	GDP deflator(%)
1975–76	333.9 (10.1)	324.0 (6.8)	3.3	382.8 (22.3)	364.2 (16.4)	5.9	208.3 (112.1)	141.6 (44.2)	67.9	1146.6 (29.0)	1025.0 (15.3)	13.7
1976–77	430.1 (28.8)	344.7 (6.4)	22.4	490.1 (28.0)	418.1 (14.8)	13.2	415.0 (99.2)	202.3 (42.8)	56.4	1468.7 (28.1)	1158.6 (13.0)	15.1
1977–78	459.3 (6.8)	327.3 (−5.0)	11.8	493.3 (0.6)	359.4 (−14.0)	14.6	453.8 (9.3)	179.3 (−11.4)	20.7	2466.4 (67.9)	1642.4 (41.7)	26.2
1978–79	543.2 (18.3)	332.4 (1.6)	16.7	437.4 (−11.3)	312.3 (−13.1)	1.8	449.3 (−1.0)	141.6 (−21.0)	20.0	2532.5 (2.7)	1558.6 (−5.1)	7.8
1979–80	767.5 (41.3)	356.4 (−7.2)	34.1	454.1 (3.8)	277.3 (−11.2)	15.0	440.9 (−1.9)	123.1 (−13.1)	11.2	2908.6 (14.9)	1634.1 (4.8)	10.1
1980–81	1054.2 (37.4)	356.6 (0.1)	37.3	502.2 (10.6)	242.3 (−12.6)	23.2	504.9 (14.5)	116.0 (−5.7)	20.2	3485.7 (19.8)	1716.1 (5.0)	14.8

Source: As Table A.5

Note:
1. Figures in parentheses refer to percentage changes.
2. The GDP deflator has been derived by subtracting the percentage change in current prices from the percentage change in constant prices.

Table A.10 Trend growth of output and price (GDP deflator) in various sectors

	1962/3 – 1972/3			1973/4 – 1978/9			1973/4 – 1980/1		
	Change in current Price(%)	Change in constant Price(%)	GDP deflator(%)	Change in current Price(%)	Change in constant Price(%)	GDP deflator(%)	Change in current Price(%)	Change in constant Price(%)	GDP deflator(%)
Agriculture	6.9 (0.99)	4.2 (0.95)	2.7	16.2 (0.97)	2.9 (0.67)	13.3	19.8 (0.96)	2.8 (0.80)	17.0
Manufacturing and mines	13.9 (0.99)	12.3 (0.99)	1.6	13.6 (0.75)	5.9 (0.21)	9.7	9.1 (0.62)	-2.1 (0.84)	11.2
Construction	13.9 (0.98)	7.6 (0.92)	6.3	39.9 (0.89)	10.9 (0.49)	29.0	27.2 (0.79)	2.2 (0.48)	25.0
Services	12.3 (0.98)	11.0 (0.94)	1.3	29.3 (0.97)	16.0 (0.94)	13.3	24.7 (0.96)	12.4 (0.90)	12.3
Wholesale and retail trade	11.6	9.0	2.6	17.1	5.1	12.0	16.0	1.5	14.5
Transport	5.7	5.1	0.6	29.0	23.3	5.7	27.6	17.5	10.1
Total (excl. oil)	11.2 (0.98)	9.1 (0.99)	2.1	23.8 (0.96)	9.8 (0.87)	14.0	20.6 (0.96)	6.9 (0.80)	13.7
Total (incl. oil)	12.3 (0.98)	11.4 (0.99)	0.9	20.0 (0.84)	3.5 (0.47)	16.5	15.8 (0.84)	-0.83 (0.30)	16.6

Source: As table A.5
Note: Figures in parentheses are values of r^2 for the trend growth regression based on the formula: $Y_t = Y_o e^{rT}$

Table A.11. Indices of wages and salaries

	Construction*	Government services*	Manufacturing	Non-oil GDP deflator*
1970–71	52.9	60.6	41.2	65.1
1971–72	54.3	60.6	47.7	68.3
1972–73	63.8	67.6	59.1	75.1
1973–74	77.6	74.0	75.4	84.0
1974–75	100.0	100.0	100.0	100.0
1975–76	147.1	117.0	143.6	113.3
1976–77	205.1	136.4	196.3	135.4
1977–78	275.6	146.1	251.4	161.0
1978–79	327.2	156.7	325.7	174.8
1979–80	377.3	179.4	527.2	198.7
1980–81	450.5	179.4	624.6	237.6
Compound growth 1973/4–1978/9	27.1	13.3	27.6	14
1973/4–1980/1	24.6	11.7	30.3	13.8

Source: As Table A.2.
* The deflator used for value added in government services.

Table A.12 Gross domestic fixed capital formation in different economic activities (billions of rials, constant 1974–5 prices)

Economic activity	1970	1971	1972	1973	1974	1975	1976	1977	1978	1978	1980
Total	286.0	334.9	410.6	463.3	529.5	874.6	1181.2	1083.1	836.0	677.8	608.8
Agriculture	17.9	25.5	34.7	36.2	46.0	59.6	58.7	46.8	31.8	35.7	31.4
Oil and gas	17.7	26.8	42.1	41.2	41.7	61.2	191.9	86.7	37.1	20.7	6.8
Manufacturing	49.9	54.0	55.0	73.5	82.6	158.0	167.1	168.1	98.3	57.4	53.7
Water and electricity	13.2	23.7	25.8	22.5	31.6	54.4	109.8	144.5	86.4	97.3	29.6
Construction	97.2	95.2	119.6	132.7	157.6	221.1	309.9	280.9	249.2	234.9	260.3
Residential	84.4	84.0	102.0	112.6	137.0	202.4	279.7	250.4	232.6	213.8	234.4
Non-residential	12.8	11.2	17.6	20.1	20.6	18.7	30.2	30.5	16.6	21.1	25.9
Services	90.1	109.7	133.4	157.2	170.0	320.3	343.8	356.1	333.2	210.6	227.0
Transport	49.2	54.6	65.2	60.9	81.1	177.8	179.7	167.2	151.7	87.2	82.7
Communication	12.4	19.9	20.4	18.5	17.5	25.7	36.3	32.6	25.0	8.4	4.9
Other	28.5	35.2	47.8	77.8	71.4	116.8	127.8	156.3	156.5	115.1	139.4

Sources: 1970–9 from *Statistical Yearbook of Iran* (1980, p. 754); 1980 from *Statistical Yearbook of Iran*, (1982–3, p. 785).

Table A.13 Sectoral distribution of GDP, (billions of rials, constant 1974 prices)

	Oil	Agriculture	Manufacturing and Mines	Construction	Services
1962–63	42.7	21.5	7.7	4.6	23.3
1963–64	43.2	20.4	7.9	5.1	23.2
1964–65	43.8	19.3	7.6	4.9	24.2
1965–66	45.2	17.9	7.5	5.5	23.7
1966–67	58.8	16.8	8.0	4.8	31.5
1967–68	47.3	16.2	8.2	5.0	23.1
1968–69	48.4	15.4	8.3	4.6	23.0
1968–70	51.1	13.9	8.0	4.1	22.6
1970–71	52.3	12.9	7.9	3.8	22.9
1971–72	54.7	10.9	8.1	3.5	22.5
1972–73	52.6	10.5	8.7	3.5	24.5
1973–74	51.3	9.9	9.1	3.5	26.0
1974–75	47.8	9.8	10.1	3.1	29.0
1975–76	41.1	10.2	11.5	4.5	32.5
1976–77	39.8	9.7	11.8	5.7	32.8
1977–78	32.9	8.7	9.6	4.7	43.8
1978–79	26.7	10.3	9.7	4.4	48.6
1979–80	22.1	11.6	9.0	4.0	53.2
1980–81	5.1	13.9	9.4	4.5	66.9

Source: Table A.10.

Table A.14 Summary Statistics, at current prices, 1962–3 (millions of rials)

Product group	Gross value added (A)	Gross output (B)	Imports CIF (C)	Exports FOB (D)	Average nominal tariffs (%) (E)	Imports at market prices (F)	Indirect taxes as % of output of 1965 (G)	Gross output at market prices (H)	Total supply at market prices (J)	Domestic demand (K)
Food	7,465	25,651	2,565	287	26.7	3,250	1.19	25,956	29,206	28,919
Beverages	723	1,066	12	0.7	69.4	20	0.44	1,070	1,090	1,089
Tobacco	4,184	6,013	1	18.8	44.4	1.4	0.17	6,023	6,024	6,005
Textiles	8,379	19,726	4,933	4,027	35.2	6,669	1.21	19,965	26,634	22,607
Apparel	1,147	4,289	151	27	109	315	0.57	4,313	4,628	4,601
Wood and furniture	797	1,982	436	58	60.9	701	5.07	2,082	2,783	2,725
Leather	513	1,648	264	442	29.2	341	0.99	1,664	2,005	1,563
Paper	252	442	1,059	4.5	125	238	13.93	504	742	737
Printing	196	463	172	0.6	0.0	172		527	699	698
Rubber	286	705	1,627	1.5	20.6	1,962	18.19	833	2,795	2,793
Chemicals	1,418	3,458	5,361	364	249	18,709	21.47	4,200	22,909	22,545
Basic metals	108	369	4,266	1*	57.3	6,710	0.11	369	7,079	7,078
Metal products	2,496	5,993	4,190	32*	63.7	6,859	21.93	7,307	14,166	14,134
Non-metallic minerals	2,814	5,242	862	5.3	25.2	1,034	2.85	5,391	6,425	6,420
Mechanical machinery	75	95	7,551	9.8	2.7	7,775	14.88	109	7,864	7,854
Electrical machinery	239	439	3,005	1.3	29.4	3,888		504	4,392	4,391
Vehicles	2,711	7,134	2,887	6.5	29.3	3,733	7.36	7,659	11,392	11,385

Source: Sadigh (1975, p.303).
F=C (1+E/100)   J=H+F
H=B (1+G/100)   K=J−D

Table A.15 Summary Statistics at current prices, 1970–71 (millions of rials)

Product group	A	B	C	D	E	F	H	J	K
Food	20,307	125,192	6,048	1,267	88	11,370	126,681	138,051	136,784
Beverages	1,575	3,428	44	7	540	282	3,453	3,735	3,728
Tobacco	7,543	8,827	10	0.3	271	37	8,842	8,879	8,879
Textiles	15,316	63,925	9,800	9,907	120	21,560	64,698	86,258	76,351
Apparel	10,510	39,138	258	703	260	929	39,361	40,290	39,587
Wood and furniture	2,723	8,172	2,167	328	228	7,108	8,586	15,694	15,366
Leather	426	3,901	307	1,526	282	1,173	3,939	5,112	3,586
Paper	856	3,635	3,666	6.2	73	6,342	4,141	10,483	10,477
Printing	1,557	5,190	371	12	73	642	5,913	6,555	6,543
Rubber	721	4,694	1,788	48.5	84	3,290	5,547	8,837	8,789
Chemicals	4,538	19,460	13,620	820	127	30,917	23,638	54,555	53,735
Basic metals	2,091	9,437	19,942	9	106	41,080	19,963	61,043	61,034
Metal products	6,182	17,553	13,975	610	63	22,779	21,402	44,181	43,571
Non-metallic minerals	9,928	14,697	2,906	160	160	7,556	15,115	22,671	22,511
Mechanical machinery	775	3,367	25,221	19	67	42,119	3,868	45,987	45,968
Electrical machinery	4,351	12,244	10,686	7	67	17,847	14,065	31,912	31,905
Vehicles	8,047	23,640	11,590	140	115	24,918	25,379	50,297	50,157

Source: see Table A.14

Note: For definitions of statistics, see Table A.14.

Table A.16 Changes in manufacturing exports, 1960–71

	Distribution 1960–1	Distribution 1970–1	Average annual growth rate (%)	Growth Share
Manufacturing exports (millions of rials)	5,653	15,571		9,918
Manufacturing exports (%)	100	100	10.5	100
Consumer goods	98.9	88.1		82.16
Food	3.6	8.1	20	10.7
Beverages	0.01	0.04	26.5	0.06
Tobacco	0.3	0.0	−50.5	−0.2
Textiles	84.2	63.6	7.3	51.9
Apparel	0.5	4.5	39	6.8
Wood and Furniture	1.1	2.1	17.5	2.7
Leather	9.2	9.8	11.5	10.2
Intermediate goods	1.0	10.7		16.23
Paper	0.0	0.04	35	0.06
Printing	0.01	0.07	34.5	0.1
Rubber	0.0	0.3		0.5
Chemicals	0.2	5.3	50.5	8.1
Basic metals	0.04	0.06	18.5	0.07
Metal products	0.7	3.9	32	5.8
Non-metallic minerals	0.09	1.0	41.5	1.6
Capital goods	0.08	1.0		1.67
Mechanical machinery	0.04	0.1	21	0.2
Electrical machinery	0.0	0.04	43	0.07
Vehicles	0.04	0.9	50.2	1.4

Source: Sadigh (1975, p. 249).

Table A.17 Average equivalent nominal duties, 1970–1

Product group	Average nominal duty (%)	value added (billions of rials)	share in total value added
Consumer goods	135*	43.01	61.8
Food	65	14.18	20.4
Beverages	485	0.67	1.0
Tobacco	246	5.50	7.9
Textile	94	13.34	19.2
Apparel	236	6.86	9.9
Wood products	195	2.18	3.1
Leather	253	0.28	0.4
Intermediate goods	95	18.56	26.7
Paper and printing	56	1.50	2.2
Rubber	66	1.54	2.2
Chemicals	116	2.99	4.3
Basic metals	87	1.36	2.0
Metal products	55	4.43	6.4
Non-metallic products	137	6.74	9.7
Capital goods	79	7.07	10.2
Machinery	59	3.08	4.4
Transport equipment	104	3.99	5.7
Other manufacturing	12	0.84	1.7
Total	–	69.57	100.0

Source: Sadigh (1975, p. 24).
* Average nominal duties were calculated by weighting the nominal duty on the sampled products by amount of the sample output. The sample in each product group was chosen from those products which carry a large weight in domestic production, and whose total share in domestic production of that product group was more than about 50%. For example the sample in textiles includes cotton fabrics, threaded cotton, carpets, woollen cloth and wool which altogether amount to about 77% of the total domestic production of textiles.
Average tariffs for main groups were obtained by weighting the average duty for each product group by the relative share of that group in total manufacturing added.

Table A.18 Ratios of average duty levels for major product groups, 1970/1960*

Product group	Duty ratio
Consumer goods	5.0
Food	6.5
Beverages	9.0
Tobacco	4.3
Textiles	3.7
Apparel	2.5
Wood products	3.7
Leather	5.3
Intermediate goods	3.9
Paper and printing	1.3
Rubber	2.3
Chemicals	1.3
Basic metals	1.0
Metal products	1.7
Non-metallic products	15.9
Capital goods	3.9
Machinery	4.4
Transport equipment	3.5
Other manufacturing	1.3
Mean	4.2

Source: As Table A.15

*Simple mean values of ratios for a selected sample of goods

Table A.19 Ex-factory and CIF prices compared for selected products, 1970–1

Product	$\frac{\text{Ex-factory}}{\text{CIF}} \times 100$	Implicit duty %	Nominal duty (%)
Consumer goods			
Food: biscuits (ordinary)	53	-47	52
biscuits (special)	65	-35	41
sugar	194	94	152
canned fruit and vegetables	98	-2	300–400
vegetable oil	120	20	101
Textiles: cotton	157	57	50–80
woollen	130	30	137–157
Wood: traverse rods	146	46	?
Intermediate goods			
Rubber: motor car tyres	111	11 ⎫	
truck tyres (USA)	51	-49 ⎬	13 to 33
truck tyres (Japan)	119	19 ⎭	
Metal products: radiators	109	9	
drawn wire	141	41	38
nails	144	44	200
cables (smallest)	108	8	38
cables (largest)	102	2	31
Chemicals: urea	90	10	?
ammonium nitrate	89	11	?
plastics (PVC)	181	81	63
caustic soda	210	110	100
Paints: decorative	126	26 ⎫	
Industrial (enamel)	142	42 ⎬	55 to 65
Industrial (high quality)	140	40 ⎭	
synthetic fibres (40 denier)	160	60	
Basic metals: rolled steel	145–140	40–45	40
Pharmaceuticals	90	-10	0
Non-metallics: glass (av.)	290	194	190
cement	41	-59	13
Paper (average quality)	137	37	21 to 98
Electrical machinery and appliances:			
refrigerators	105	5	60
radios	120	20	100
TVs	156	56	100
space heaters	114	14	?

Table A.19 Cont.

Product	$\frac{\text{Ex-factory}}{\text{CIF}} \times 100$	Implicit duty %	Nominal duty (%)
Capital and durable consumer goods			
coolers	200	100	35
car batteries	203	103	16
bulbs	160	30	
transformers	110	10	15
electric fans	178	78	45
electric switchgear	85	-15	?
telephones	110	10	?
Transport equipment:			
Hillman	148	48 ⎫	
buses	128	28 ⎬	225–300
Rambler	113	13 ⎮	
Citroën	194	94 ⎭	
trucks (Leyland)	173	73 ⎫	20–40
trucks (Merecdes)	189	89 ⎭	
Mean	135	35	84

Source: Sadigh (1975, pp. 64–5).

Table A.20 Iranian industry: import dependence

	Share of imported inputs in sales value (%)	Ratio of ex-factory to CIF price (%)
Light consumer goods		
Sugar	2	194
Meat packing	10	
Cotton textiles	10	129, 97
Footwear	10-20	<100
Canned fruits and vegetables	10-20	112, 93
Woollen textiles	40	97
Vegetable oils	50	97
Pharmaceuticals	60	<100
Durable consumer goods		
Electric fans†	26	
Radios	37-50	120
Space heaters	40	86
Refrigerators	40 (understate?)	152, 133, 79
Air coolers	60	
TVs	60	108, 101
Transport equipment		
Diesel engines†	33-43	
Trucks	48	137, 107
Buses	n.a. (probably as trucks)	106
Passenger cars	50	115
Tyres	50	105, 90
Intermediate products		
DDB (Docedil Benzane)†	10	
PVC†	10	
Caustic soda	10	
Glass (sheet)	10	105, 69
Paper†	25	
Paints	45	
Synthetic fibres	n.a. (above 50?)	>100
Rolled steel	60	

Table A.20 Cont.

	Share of imported inputs in sales value (%)	Ratio of ex-factory to CIF price (%)
Capital goods		
Cement	20	44
Carbon steel and stainless blades	20	
Telephone receivers	20	
Electric meters†	20	
Telephone exchanges†	20	
Steel wire, nails and screws	30 (?)	Low
Pumps	40	
Transformers	45	
Cables	65-90	
Electric switchgear	80	

*Imported components include raw materials, semi-finished products and spare parts. Both direct and indirect import content are included (The indirect content being defined as imported materials bought on the home market rather than directly imported by the user).
†Project under construction; planned values.

Table A.21 Quantities of output and import (export) of selected industries

	1972–3		1973–4		1974–5		1975–6		1976–7		1977–8		Price growth rate 1972–7 (%)
	Output	Imports (Exports)	Output	Imports (Exports)	Output	Imports (Exports)	Output	Imports (Exports)	Output	Imports (Exports)	Output	Imports (Exports)	
Cotton/synthetic fibres (millions of meters)	495	85			600		635		510	240	450	210	-1.9
Footwear (millions of units)	80	(10)			89.5	(6)	110		125	(3)	128	(3.5)	+9.9
Canned fruits & vegetables (thousands of tonnes)	20.7				25		29.9			33.5	54		+21
Cement (millions of tonnes)	3.3	0.07			4.7	(0.73)	5.4	1.1	6.1	1.2	7	1.7	+16.2
Bricks (millions of units)	6,000				8,000				13,900		13,400		+17.4
Paper (thousands of tons)	21.3		31	37					46		54		+20.4
Woollen textiles (millions of metres)	13			16			18		16.5	16	19.7	12	+8.7
Vegetable oil (thousands of tonnes)	188			244			260		300		315		+10.8
Motor-car tyres (thousands of tonnes)	20.1			26			33		37.9		44.5		+17.2

	1972–3		1973–4		1974–5		1975–6		1976–7		1977–8		Price growth rate 1972–7 (%)
	Output	Imports (Exports)	Output	Imports (Exports)	Output	Imports (Exports)	Output	Imports (Exports)	Output	Imports (Exports)	Output	Imports (Exports)	
Sugar (thousands of tonnes)	667		700				150	250	690		630		−1.1
Sheet glass (thousands of tonnes)	46	11	60		64		9.3		80		63		+6.4
Refrigerators (thousands of units)	185		350				440				570	+25	
Televisions (thousands of units)	170		300				370				205	+3.8	
Passenger cars (thousands of units)	49		72				87				139.8	+23.3	

Declining industries	Slow growth <10%	Fast growth >10%
1. Cotton textiles	1. Television	1. Refrigerators 5. Bricks, Cement
2. Woollen textiles	2. Sheet glass	2. Canned fruits and vegetables 6. Tire of automobiles
3. Sugar	3. Footwear (marginal)	3. Passenger cars 7. Vegetable oil
		4. Paper

Source: As Table A.15.

Notes:
1. Question mark indicates that there are other conflicting estimates; or the data are a guess; or the data are unreliable; or the data do not seem plausible.
2. In some cases only one plant producing a particular product was interviewed; as a result, the data may not be representative of the situation of the industry. This qualification does not hold when the product is produced by one plant only: there are a number of such cases. Furthermore, when more than one plant was interviewed, the cost composition and the cost levels appeared remarkably similar.

Source: Interviews with the industrial plants and material in IMDBI files. Cited in Avramovic (1970).

Table A.22 Import price and ex-factory price for selected manufactured goods.

SITC		1977 CIF import price(A)	1979 Ex-factory price(B)	A/B
0612	Refined sugar* c	30.1	51.56	1.71
6512	Yarn of wool* i	702.1	513.9	0.73
6516	Synthetic yarn i	245.6	318.9	1.29
641	Paper and paperboard ci	34.8	51.29 (paper) 26.4 (paper-board) 38.8 (average)	1.11
6612	Cement k	4.53	2.6	0.57
75201	Non-domestic requ. i	400	258	0.64
7241	Television c	14.2	21.6 (colour) 79.4 (black and white)	5.59 (colour) 1.52 (black and white)
7321	Cable i	22.3	20.6	0.92

* For sugar and wool CIF prices have been updated by using the percentage increase in price of sugar and wool in major commodity markets (New York), from IMF (1982, pp.89–95).
c consumer goods
i intermediate goods
k capital goods

Source: CIF import prices have been calculated by dividing price by quantity from the UN *International Trade Statistic Yearbook* 1981; domestic ex-factory prices are calculated in the same way from *Statistical Yearbook of Iran* (1980, pp. 472–5).

Table A.23 IMDBI financial assistance by activity

	1971–2	1972–3	1973–4	1974–5	1975–6	1976–7	1977–8
Agriculture and industry	–	–		–	0.5	–	–
Mining and Quarrying	–	–		0.1	–	4.1	–
Industry							
Food processing	–	10.3		11.1	13.7	28.7	41.8
Textiles	0.2	14.2		8.8	4.1	11.3	25.3
Footwear	–	–		–	–	–	–
Wood industries (except furniture)	–	–		–	0.3	0.1	0.1
Furniture	–	–		–	0.4	–	1.4
Paper and paper products	13.2	7.6		0.1	–	1.5	0.3
Printing	–	–		–	0.1	1.6	5.2
Leather and leather products	–	–		–	2.2	0.2	0.1
Rubber products	–	–		0.2	–	–	–
Chemical and chemical products	1.4	1.9		0.4	–	0.7	3.8
Petroleum products	–	–		–	–	–	–
Non-metallic minerals	7.6	5.1		7.3	0.8	1.7	3.0
Metal products	49.1	21.0		14.5	6.4	10.4	5.3
Industrial machinery	–	1.5		–	0.4	–	3.1
Electrical machinery	3.0	2.2		3.1	1.3	1.2	4.2
Transport equipment	23.0	0.7		2.0	–	1.4	2.7
Miscellaneous	2.3	1.7		4.2	–	–	0.9
Construction activities	–	–		1.4	0.1	3.0	–
Commercial enterprises	–	0.2		–	–	–	–
Transport, storage, communications	–	–		–	–	0.3	–
Services	–	33.6		6.9	69.7	5.1	2.4
Banking	–			39.4		28.7	0.4
Total	100	100		100	100	100	100

Source: IMDBI (1972–8).

Table A.24 Production index of large manufacturing establishments (1974–5 = 100)

	1975–6	1976–7	1977–8	1978–9	1979–80	1980–1
Food, beverages and tobacco	108.4	122.8	123.4	111.9	117.2	105.6
Textiles, clothing and leather	116.8	130.6	149.8	143.7	162.8	172.5
Wood and wooden products	119.5	143.2	191.5	201.0	219.9	210.2
Paper, cardboard and their products	105.2	125.0	148.9	134.7	135.2	96.0
Chemicals	108.2	127.4	148.1	119.2	118.8	98.3
Non-metal mining products (except oil and coal)	119.7	147.8	162.8	151.3	177.7	181.7
Basic metal	108.0	112.5	143.1	113.7	104.8	83.4
Metal machinery and equipment	125.2	158.0	175.6	138.8	118.8	113.4
Other manufacturing industries	82.2	92.0	98.5	70.6	58.7	29.4
General index	114.7	134.6	150.6	129.1	129.7	121.5

Source: Bank Marzaki Iran, *Annual Report and Balance Sheet* 1977–8; 1980–1.

Table A.25 Share of different groups of manufactured goods in total manufactured output (per cent)

	1971	1976	1979
Food	26	21	21
Textiles	18	14	16
Paper	2	2	2
Chemicals	12	22	16
Construction related	4	6	9
Basic metals	15	7	8
Machinery	19	24	22
Other	4	4	6

Source: Statistical Yearbook of Iran (1980, p. 469).

Table A.26 Agricultural output by farm sizes

Farm size (ha)	Families or farms '000	(%)	Farm area (ha) '000	(%)	Average size (ha)	Share of gross output	Share of marketed output
Large Farms (>100)	7	–*	1,810	12	258	6 ⎫	⎫
Medium sized farms						⎬	⎬ 77
51–100	10	–†	700	4	70 ⎫	⎬ 36 ⎭	⎭
11–50	394	16	7,030	46	18 ⎭		
Small Farms							
6–10	434	17	3,180	21	7.3 ⎫	⎫	⎫
3–5	545	22	1,810	12	3.3 ⎬	⎬ 41	⎬ 19
1–2	342	13	490	3	1.4 ⎬	⎭	⎭
<1	801	32	300	2	0.4 ⎭		
Sub-total	2,533	100	15,330	100	6.0	83	
Pastoralists	100				⎫	⎫	⎫
Other people not owning land	790				⎬	⎬ 17	⎬ 4
					⎭	⎭	⎭
Total	3,423		15,330			100	100

Source: Price (1975).
* less than 0.002 per cent
† less than 0.004 per cent

Table A.27 Membership and provision of loans by the rural co-operatives

	Membership (000)	Total Loans (billions of rials)	Loan recipients (000)	Members receiving loans (%)	Average loan size (rials)
1963–64	571	0.50	151	26	3,329
1964–65	668	1.43	328	49	4,355
1965–66	791	1.88	391	49	4,814
1966–67	961	3.02	558	58	5,411
1967–68	1,105	4.07	670	61	6,080
1968–69	1,278	5.04	738	58	6,826
1969–70	1,430	5.75	833	58	6,899
1970–71	1,606	6.31	901	56	7,002
1971–72	1,723	6.81	876	51	8,015
1972–73	2,065	10.07	1165	56	8,650
1973–74	2,263	12.37	1176	52	5,470
1974–75	2,488	19.74	1363	55	14,482
1975–76	2,685	24.70	1611	60	15,330
1976–77	2,868	28.00	1400	49	20,000
1977–78	3,001	30.10	1320	44	22,800

Sources: Ministry of Rural Affairs and Cooperatives, *Rural Cooperatives in Iran*, Research Group no. 33, Tehran, 1971; Bank Marzaki Iran, *Annual Report*, 1973/4–1978/9.

Table A.28 Provision of credit by the Agricultural Development Bank of Iran

	1971–2	1972–3	1973–4	1974–5	1975–6	1976–7
Number of loan applications approved	69	159	309	534	682	580
Amount of credit and outright aid approved (millions of rials)	1,301	1,282	2,560	17,205	48,600	25,801
Average credit and aid given per project (millions of rials)	18.8	8.0	18.2	32.2	71.2	44.4
Average total investment per project (millions of rials)	43.3	30.8	18.2	57.7	148.8	91.5
Percentage of total investment	43.5	26.1	45.4	55.7	47.9	48.5

Source: Agricultural Development Bank of Iran, *Annual Report*, 1972–7

Table A.29 Agricultural Production by Commodity (thousands of tonnes)

	Wheat	Barley	Paddy* rice	Sugar-cane	Sugar-beets	Cotton
1968/69	4,194	1,168	(1,000)	495	3,412	545
1969/70	4,209	1,143	(1,040)	514	3,484	517
1970/71	4,004	1,083	(1,138)	509	3,455	513
1971/72	3,612	851	(1,046)	554	3,772	459
1972/73	4,398	1,227	1,015	559	3,639	630
1973/74	4,546	1,158	1,023	609	3,423	615
1974/75	2,886	751	1,500	640	4,075	715
1975/76	4,366	1,019	1,500	940	4,670	560
1976/77	–	–	1,600	800	5,250	426
1977/78	3,896	1,130	1,400	1,000	4,150	531
1978/79	3,791	1,133	1,531	1,700	3,900	427
1979/80	4,570	1,411	1,271	1,500	4,900	322
1980/81	3,733	979	1,181	1,400	–	220

Source:Statistical Yearbook of Iran (1973–4, 1979–80, 1980–1).

* Centre for Agricultural Marketing Development, Rice Production and Marketing in Iran, Tehran, 1977, p.12.

* 1968–75 from A. Mojtahedi, *Rice Growing in Northern Iran*, Occasional Publications no.15, Dept. of Geography, University of Durham, 1980, p.7; 1975–79 from Bank Marzaki Iran, Annual Report and Balance Sheet, 1979–80, p.141; 1982–83, p.142.

Table A.30 Imports of agricultural commodities (thousands of tonnes)

	Wheat	Barley	Sugar-beet	Cotton (D) Imports	Cotton (D) Exports
1968–69	535,500	0	0	0	77,378
1969–70	700	1	0	0	91,106
1970–71	23,300	99	0	0	108,168
1971–72	997,800	191,862	0	0	102,401
1972–73	774,300	23,145	0	6	115,556
1973–74	875,700	107,524	2,487	22	124,630
1974–75	1,449,800	178,483	0	5	86,685
1975–76	1,463,500	203,864	0	9	145,870
1976–77	432,700	219,784	0	0	92,869
1977–78	1,274,700	333,806	0	1,384	63,800
1978–79	1,056,000	337,371	0	85	62,591
1979–80	1,060,000	122,850	0	7	56,327
1980–81		387,985	0	2	3,471
1981–82		471,998	0	481	0

Source: FAO *Trade Yearbook*; wheat figures from Bank Marzaki Iran, *Annual Report*, various years.

Table A.31 Government support price of selected crops (rials per tonne)

	Wheat	Barley	Soybeans	Sunflower seeds
1972–73	6,000			
1973–74	6,000			
1974–75	10,000	4,000		
1975–76	10,000	7,500	24,000	20,000
1976–77	10,000	7,500	24,000	20,000
1977–78	12,000	7,500	24,000	20,000
1978–79	14,000	7,500	24,000	20,000
1979–80	18,000	16,000	26,000	22,000
1980–81	20,000		40,000	45,000
1981–82	36,000			

Source: FAO (1983b).

Table A.32 Domestic and import price of wheat

	Average domestic wholesale price of wheat (rials/kg)	Average import price CIF (rials/kg)	Ratio of domestic to import price
1962–63	7.86	5.44	1.44
1963–64	7.41	8.87	0.83
1964–65	7.95	5.91	1.34
1965–66	8.33	4.95	1.68
1966–67	7.38	5.41	1.36
1967–68	5.98	5.32	1.12
1968–69	5.52	5.26	1.04
1969–70	6.14	16.47*	.37
1970–71	7.68	7.38	1.04
1971–72	8.40	5.79	1.45
1972–73	7.11	6.16	1.15
1973–74	7.92	6.45	1.22
1974–75	12.58	14.73	.85
1975–76	13.40	15.75	.85
1976–77	n.a.	17.62	—
1977–78	12.60	11.37	1.10

Sources: O. Aresvik, Agricultural Development in Iran, Praeger Publishers, New York, 1976. Foreign Trade Statistics of Iran, Annual Report, Ministry of Commerce, Tehran, various years.

* Only 499 tonnes were imported in that year.

Table A.33 Annual average wholesale price of different varieties of rice on Rasht market (rials per kilogram).

	Domsiah		Sadri Dargi (1st clss)		Sadri Machini (1st class)		Gharib		Champa	
	Price	Index	Price	Index	Price	Index	Price	Index	Price	Index
1971–72			31	57	26.6	63	24.2	64	19	62
1972–73			33	61	26	61	24.8	66	14.5	47
1973–74			38	70	30.2	71	29.3	78	18.4	60
1974–75	60.4	100	54	100	42.2	100	37.3	100	30.5	100
1975–76	59.5	98	51	94	32.7	77	28.8	77	19.3	63
1976–77	76.2	128	69	127	45.3	107	36.9	98	24.4	80
1977–78	87.4	144	76	140	48.3	114	46.6	124	30.5	100
1978–79	91.4*	151								
1979–80	135.3*	224								
1980–81	139.0*	230								

Source: Centre for Agricultural Marketing Development, 'Rice Production and Marketing in Iran', pp. 61–7.

* Due to the absence of information, we have used the retail price of rice in rural areas.

Table A.34 Nominal protection coefficient for rice

Year	Unit value of imported rice CIF ($)	Unit import value (rials/kg)	Ratio of domestic wholesale price to the import price for different grades of rice*					Price of rice in New Orleans ($/tonne)
			I	II	III	IV	V	
1970	198.3	15.0						189.6
1971	192.4	14.5						191.8
1972	224.0	16.9		2.1	1.8	1.6	1.3	216.0
1973	269.0	18.5		1.9	1.5	1.4	0.8	316.8
1974	851.3	57.5	1.0	2.0	1.6	1.5	0.9	555.5
1975	575.3	38.9	1.5	0.9	0.7	0.6	0.5	418.8
1976	451.6	31.7	2.4	1.3	0.8	0.7	0.5	308.6
1977	475.0	33.5	2.6	2.1	1.4	1.1	0.7	332.8
1978	547.6	38.5	2.4	2.2	1.4	1.3	0.9	399.0
1979	511.3	36.0	3.8					381.4
1980	519.4	36.6	3.9					496.4

Sources: Calculated from the FAO *Trade Yearbook*, various years; New Orleans prices from IMF (1983, p. 91).

*Calculated by dividing unit import values (rials/kg) by the appropriate price from Table A.33.

Table A.35 Production and imports of sugar (thousands of tonnes)

	Domestic output	Imports	Imports ×100 / Domestic output
1967–68	415.3	202.7	48.8
1968–69	485.1	34.1	7.0
1969–70	512.1	66.0	12.9
1970–71	568.8	61.4	10.8
1971–72	585.7	87.5	10.1
1972–73	579.6	158.8	27.4
1973–74	608.5	286.0	47.0
1974–75	639.5	219.5	47.0
1975–76	665.9	596.4	89.7
1976–77	721.9	262.4	36.3
1977–78	632.3	447.5	70.7

Source: Cereals, Tea and Sugar Organisation, *Annual Report on Output of Sugar Factories*, Tehran, various years; Foreign Trade Statistics of Iran, *Annual Report*, Ministry of Commerce, Tehran, various years. Quoted in Madjd (1983).

Table A.36 Domestic and import price of granular sugar (rials per kilogram)

	Price paid to domestic factories	Average import price (CIF)
1971–72	15.75	9.06
1972–73	15.75	12.52
1973–74	19.25	17.85
1974–75	25.50	43.70
1975–76	32.00	61.25
1976–77	35.50	64.56
1977–78	39.45	25.32
1978–79	43.85	16.09

Source: M. G. Madjd, '*Instability of Sugar Production in Iran*', Plan and Budget Organisation, Tehran, 1981 (in Persian). Quoted in Madjd (1983).

Table A.37 Comparison of sugar-beet output, prices, and labour costs

	Total output (millions of tonnes)	Index of sugar-beet prices	Index of urban construction labour costs
1969–70	3.48	59.2	43.4
1970–71	3.85	59.3	51.9
1971–72	3.96	61.5	55.4
1972–73	3.92	63.8	65.0
1073–74	4.23	71.7	75.8
1974–75	4.12	100.0	100.0
1975–76	4.59	136.9	151.4
1976–77	5.27	152.1	204.3
1977–78	4.18	162.3	273.2

Sources: M.G. Madjd, '*Policies Concerning Sugar Production in Iran*', Ph.D.Thesis, Cornell University, Ithaca, NY, 1978; Bank MarzakiIran, *Annual Report*, various years. Quoted in Madjd (1983).

Table A.38 Production and imports of some non-cereal crops (thousands of tonnes)

Crop	1970	1971	1972	1973	1974	1975	1976	1977	1978	1979	1980	Average	Cumulative growth (%)
Tea	80	64	88	93	96	80	88	na	na	34	43	(1970–76) 84.0	–46
Change on previous year (%)		–20	37.5	5.7	3.2	–17	10	na	na	na	26		
Imports (Level relative to the average)	–	–	+	+	+	–	+	+	+	+	–		
Imports	6.2	7.3	9.0	7.5	12.5	12.3	13.5	18.8	19.7	20.0	22.0		
Oil seeds	58	46	54	57	79	100	130	105	126	99	70	(1970–80) 84	21
Change on previous year (%)	–21	17.4	5.6	39	26.6	30	–19.2	(20)	(–21)	(–29.2)			
Imports (Level relative to the average)				–	–	+	+	+	+	+	–		
Imports													
Tobacco	na	19	24	15	14	15	19	15	16	20	24	(1971–80) 18.0	26
Change on previous year (%)			(26.3)	(–37.5)	(–6.7)	(7.1)	(27)	(–21)	(6.7)	(25)	(20)		
Imports (Level relative to the average)			+	–	–	–	+	–	–	+	+		
Imports	ng	ng	ng	ng	–	–	+	–	–	+	+		
										1.0	1.0		
Pulses	205	196	176	200	210	225	220	187	188	227	227	(1970–80) 205	11
Change on previous year (%)		(4.4)	(10.2)	(13.6)	(5.0)	(7.1)	(–2.2)	(–15)	(0.5)	(20.7)	(0)		

Table A.38 (cont)

Crop	1970	1971	1972	1973	1974	1975	1976	1977	1978	1979	1980	Average	Cumulative growth %
(Level relative to the average)	0	–	–	–	+	+	+	–	–	+	+		
Imports	ng	0.1	7.2	7.3	6.9	8.3	8.0	26.0	15.0	13.0	10.0		
Potatoes	na	400	420	481	533	550	570	697	932	997	1,131	(1971–80) 671	183%
Change on previous year (%)	na	na	(5)	(14.5)	(10.8)	(3.2)	(3.6)	(22.3)	(33.7)	(7.0)	(13.4)		
(Level relative to the average)	na	–	–	–	–	–	–	+	+	+	+		
Imports	0	0	0	0	0	7.0	5.2	12.5	na	13.0	27.0		
Onions	na	250	258	307	305	330	340	392	502	515	638	(1971–80) 383	155%
Change on previous year (%)	na	na	(3.2)	(9.0)	(–0.7)	(8.2)	(3.0)	(15.2)	(28)	(2.6)	(23.9)		
(Level relative to the average)	na	–	–	–	–	–	–	+	+	+	+		
Imports	0	0	0	0	3.9	8.9	55.0	0	0	0	0		
Tomatoes	na	na	130	126	285	282	372	na	na	na	na	(1972–76) 237	186%
Change on previous year (%)	na	na	na	(–3.1)	(126)	(–1)	(3)	na	na	na	na		
(Level relative to the average)	na	na	–	–	+	+	+	na	na	na	na		
Imports	0	0	0	0	0	0	0	0	0	0	0		

Sources: Ministry of Agriculture and Rural Development as reported in the Bank Marzaki, Iran, *Annual Report and Balance Sheet*, 1971–81. Also quoted in 'The Rural and Agricultural Statistics of Iran' in *The Land and the Peasant Problems in Iran*, Agah Publishers, Tehran, 1982. The figures for imports are taken from the FAO Trade Yearbooks 1973, 1976, 1979, 1981.

Table A.39 Distribution of credit by economic sector (billions of rials)

	Domestic trade	Manufacturing and mines	Construction	Imports	Exports	Agriculture	Others	Total
1968–9	42.8	25.2	22.9	18.3	6.0	18.9	30.9	165.0
	(25.9)	(15.3)	(13.9)	(11.1)	(3.6)	(11.5)	(18.7)	(100.0)
1969–70	51.2	25.2	26.3	19.5	10.2	20.3	37.7	190.4
	(26.9)	(13.2)	(13.8)	(10.2)	(5.4)	(10.7)	(19.8)	(100.0)
1970–1	57.8	30.5	29.0	19.4	12.1	20.3	57.2	226.3
	(25.5)	(13.5)	(12.8)	(8.6)	(5.3)	(9.0)	(25.3)	(100.0)
1971–2	73.6	34.6	34.4	23.5	14.5	18.4	68.9	267.9
	(27.5)	(12.9)	(12.8)	(8.8)	(5.4)	(6.9)	(25.7)	(100.0)
1972–3	104.3	42.6	45.8	31.1	17.8	17.4	68.0	327.0
	(31.9)	(13.0)	(14.0)	(9.5)	(5.5)	(5.3)	(20.8)	(100.0)
1973–4	139.7	90.8	61.7	57.8	23.2	37.3	76.2	486.7
	(28.7)	(18.7)	(12.7)	(11.9)	(4.8)	(7.6)	(15.6)	(100.0)
1974–5	173.8	144.1	85.4	112.0	22.3	62.6	91.7	691.9
	(25.1)	(20.8)	(12.3)	(16.2)	(3.2)	(9.1)	(13.3)	(100.0)
1975–6	271.0	190.3	147.8	136.8	38.0	94.0	190.6	1,068.5
	(25.4)	(17.8)	(13.8)	(12.8)	(3.6)	(8.8)	(17.8)	(100.0)
1976–7	361.9	369.2	237.3	167.0	21.9	130.0	177.3	1,464.6
	(24.7)	(25.2)	(16.2)	(11.4)	(1.5)	(8.9)	(12.1)	(100.0)

Source: Bank Marzaki Iran, Annual Report and Balance Sheet, 1971–2, 1974–5, 1977–8.

Note: Figures in parentheses refer to percentage shares.

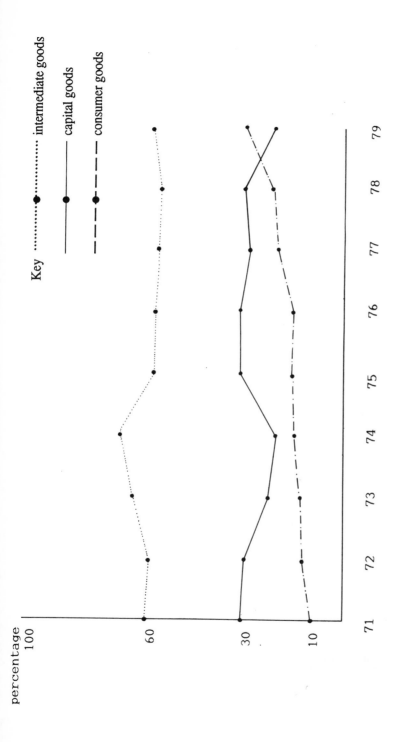

Figure A.1 Shares of different components of total merchandise imports

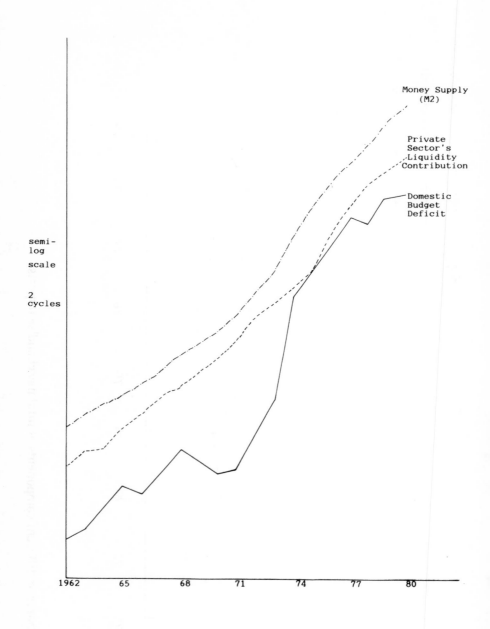

Figure A.2 Money supply and its determinants, Iran, 1960–80

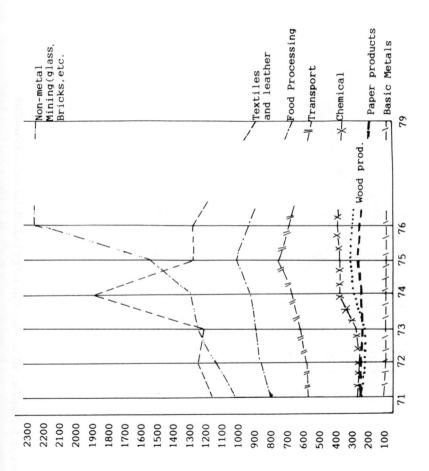

Figure A.3 Number of large industrial establishments

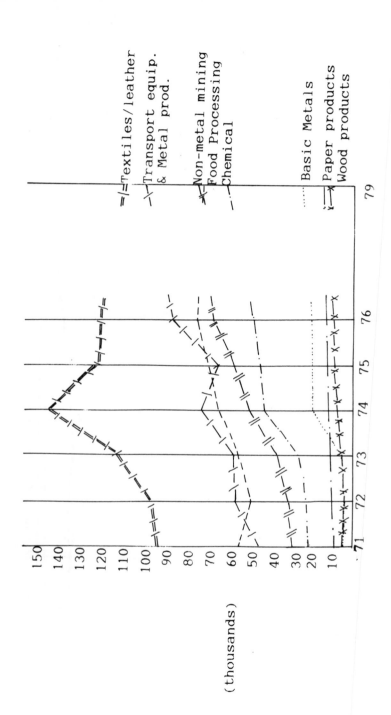

Figure A.4 Number of workers in large industrial establishments

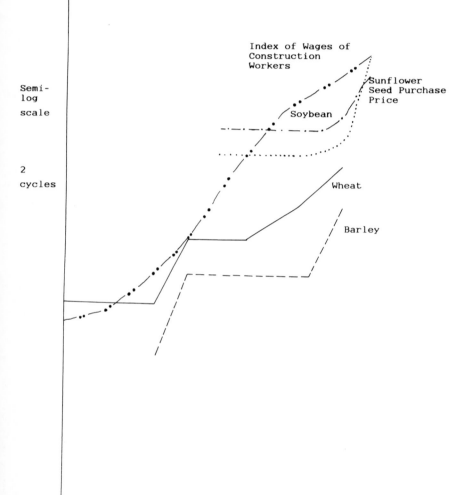

STATISTICAL APPENDIX B: NIGERIA

Table B.1 Oil exports (millions of rials)

	Total exports	Oil exports	(%)
1970	886	509.6	57.5
1971	1,303	962.8	73.9
1972	1,433	1,174.8	82.0
1973	2,319	1,934.7	82.0
1974	6,095	5,665.3	92.9
1975	4,791	4,592.1	95.8
1976	6,322	5,895.0	93.2
1977	7,594	7,045.8	92.8
1978	6,707	6,033.4	90.0
1979	10,549	10,034.6	95.1
1980	14,513	13,998.6	96.5

Source: IMF (1983).

Table B.2 Federal government revenues (millions of rials)

	1970	1971	1972	1973	1974	1975	1976	1977	1978	1979
Direct taxes	144.4	451.2	624.4	852.0	3,032.1	2,990.2	3,852.4	4,839.8	3,962.3	5,753.8
Personal income tax	0.8	–	–	1.2	11.1	15.9	3.5	3.3	3.3	2.9
Company income tax	45.8	63.0	80.4	75.5	146.6	261.9	222.2	476.9	527.4	575.1
Petroleum profit tax (A)	97.6	383.2	540.5	769.2	2,872.5	2,707.5	3,624.9	4,330.8	3,415.7	5,164.2
Other tax revenues	0.2	5.0	3.5	7.0	1.9	4.9	1.8	28.8	15.9	11.6
Indirect taxes	369.4	491.1	481.1	516.2	498.2	760.7	882.7	1,145.6	1,698.3	1,144.0
Import duties	215.6	284.8	274.4	307.9	328.3	629.3	724.3	964.2	1,436.3	870.7
Export duties	41.2	37.8	26.9	12.5	5.5	5.8	6.1	4.2	2.8	0.2
Excise duties	112.6	168.6	179.8	196.0	164.4	125.5	152.3	266.1	259.2	273.1
Interest and repayments	26.4	36.6	44.8	49.8	127.1	162.7	189.0	–	523.6	274.6
Mining (rent, royalties) (B)	68.8	127.0	223.8	246.8	854.2	1,564.0	1,740.3	1,749.8	1,238.4	3,716.7
Miscellaneous	24.2	63.0	30.7	29.6	25.4	37.1	101.5	41.1	46.8	23.6
Total (C)	633.2		54.4	59.9	82.1	77.5	79.3	75.6	62.3	81.4
% of oil in Total revenues $(100[(A+B)/C])$	26.3	43.6	54.4	59.9	82.1	77.5	79.3	75.6	62.3	81.4

Source: Federal Office of Statistics, *Annual Abstract of Statistics*, 1981.

Table B.3 Imports, total payments and exports (millions of dollars)

	Merchandise imports	Total current payments (A)	Total exports (B)	A/B
1970	939	1,772	1,341	1.32
1971	1,393	2,418	2,010	1.20
1972	1,366	2,636	2,316	1.14
1973	1,714	3,718	3,763	0.99
1974	2,480	5,052	10,048	0.50
1975	5,484	8,961	9,130	0.98
1976	7,478	11,125	10,924	1.02
1977	9,723	14,184	13,349	1.06
1978	11,685	15,197	11,684	1.30
1979	10,512	14,541	19,309	0.75
1980	15,830	21,751	27,301	0.80
Average growth rate (1973–80) (%)	32.0	24.7	28.1	

Source: IMF (1983).

Table B.4 Change in nominal rate of protection (per cent)

	1957	1962	1967	1977	1978	1979	1967–77	1977–79
Consumer goods	33.56	53.72	80.84					
Meat products	0	50	51.8	0	0	10	−100	—
Misc. food products	25	50	75					
Bakery products	50	33.3	66.6	40	75	40	−40	0
Alcoholic drinks	75	166.0	120.0	27–58	72	25–75	−77, −51	0, +2
Soft drinks	0	50	75	22	22–75	20–35	−72	0, +3
Tobacco products	50	66.6	100.0	45	40	45	−55	0
Footwear	25	33.3	40.0					
Clothing and made-up textiles	25	33.3	50.0	23–34	40	45	−54, −32	+48, +2
Wood products	0	50	50	90	33,55, 100	90	+80	0
Furniture	20	33.3	75	33–55	33–55			
Cosmetics and perfume	75	160	100				−56, −26	0
Soap and soap products	20	33.3	50	41	30	26–50	−18	−36, +2
Matches	25	33.3	66.6					
Musical instruments	33.3	50	66.6					
Clocks and watches	20.0	33.3	50					
Radios and radiograms	0	50	66.6	45	70	70	−32	+35
Cameras and projectors	33.0	66.6	75					
Printing and publishing	16.0	10	10		0			
Intermedite goods	17.50	25.98	41.25					
paper and paper products	15	20	25					
Leather products	10	33.3	75.0					

	1957	1962	1967	1977	1978	1979	1967-77	1977-79
Rubber processing	15	21.4	66.6	40	40		-26	
Tyres and tubes	0	33.3	33.3	22		20-30	-66	0,+3
Industrial chemicals	0	33.0	33.3	35	10	10	+0.6	-71
Prints and varnishes	20	33.3	33.3	28	33	25-33	-33	-10,+1
Petroleum and coal products	20	30.0	33.3					
Ceramics and tiles	0	33.3	50.0	20-33	20-33		-60,-34	
Plastic goods	0	33.3	66.67	23	20		-65	0,+1
Concrete products and cement	18.67	20.0	33.3	18,0	6,0	18, 109	-45	
Textiles	20.0	35.0	40.0	14-40			-26	
Yarns and threads	0	20.0	33.3	22-45			-33.1, +35	
Refrigeration equipment	0	33.3	40.0	40	40			
Capital goods	8.42	21.41	43.75					
Electrical equipment	0	33.3	40.0	35	75	35-70	-12.5	0, +10
Office machinery	5	25.0	40.0					
Glass products	0	33.3	33.3	28		35-40	-15.9	+25, +42
Aluminium products	0	10.0	10.0					
Copper products	0	10.0	10.0					
Iron & steel products	0	10.0	10.0	45	15		+450	
Metal wire	0	10.0	10.0					
Misc. metal products	5	10.0	12.0					
Non-electrical machinery	0	10.0	10.0	11		5	+10	-54
Bicycle and assembly	10	25.0	33.3	15	25	25	-54	+66
Motor-vehicle assembly	20	30.0	75.0	25-100	50-500	50-500	-66, +33	+100, +5

Source: 1957, 1962, 1967 from T.A. Oyejide, *Tariff Policy and Industrialisation in Nigeria*, Ibadan University Press, Ibadan, 1975, pp.52–3; 1977 and 1978 from James W. Robertson, 'The Structure of Industrial Incentives in Nigeria 1978–79', World Bank, 11 November 1980; 1979 from J.W.Robertson, Industrial Incentives in Nigeria 1979–80' World Bank, 15 September 1981.

Table B.5 Money supply and domestic deficit (millions of naira)

	Money Supply (M_2)	Domestic deficit	Claims on the private sector
1973	1,508	487	847
1974	2,730	1,203	1,121
1975	4,178	3,145	1,750
1976	5,843	3,646	2,382
1977	7,813	4,194	3,459
1978	7,521	3,228	4,485
1979	9,849	6,025	5,126
1980	14,390	8,016	6,744

Sources: IMF (1983); Domestic deficit figures World Bank, *Nigeria: Country Economic Memorandum*, 1981.

Table B.6a Average annual growth rate of expenditure categories of GDP (per cent)

	Current (1973–8)	Constant (1970–80)
Private consumption expenditure	20.4	6.6
Gross fixed capital formation	34.8	15.8
Government consumption expenditure	28.2	11.3
GDP at market prices	21.3	6.5

Sources: World Bank, *World Development Report*, 1982; IMF (1983).

Table B.6b Distribution of expenditure categories of GDP

	1970	1971	1972	1973	1974	1975	1976	1977	1978
Private consumption expenditure	73.7	71.7	68.4	66.9	53.7	60.7	60.0	58.9	63.6
Gross fixed capital formation	15.7	18.1	18.2	16.7	19.0	24.2	24.9	29.4	31.4
Government consumption expenditure	10.2	8.9	10.4	9.1	6.6	12.9	12.1	12.8	12.6
Total	99.6	98.7	97.0	92.7	79.3	97.8	97.0	101.1	107.6

Sources: Calculated from IMF (1983).

Table B.7 Federal government revenue and expenditure (millions of naira)

Fiscal	Total federally collected Revenue	Total federal expenditures	Total current expenditure	Total capital expenditure	Index of current expenditure 1972/3 = 100	Index of capital expenditure 1972/3 = 100	Index of total federal expenditure 1972/3 = 100
1971–72	1,169.0	2,632.4	918.6	173.8	141	65	38
1972–73	1,404.8	1,863.7	1,412.4	451.3	100	100	100
1973–74	1,695.3	3,642.5	1,213.1	565.7	195	85	125
1974–75	4,357.0	4,260.3	2,710.9	1,549.4	229	191	343
1975–76	5,514.7	8,258.3	4,740.1	3,518.2	443	335	779
1976–77	6,795.9	9,701.5	5,459.6	4,241.9	520	386	940
1977–78	8,042.4	11,695.3	6,253.0	5,442.3	627	443	1,205
1978–79	7,459.4	12,337.1	7,140.1	5,197.0	662	505	1,151
1979–80	10,912.7	13,191.5	8,354.0	4,837.5	707	591	1,072
1980–81	15,234.0	23,695.7	15,300.1	8,395.6	1,271	1,083	2,081

Source: Central Bank of Nigeria, *Economic and Financial Review* 1976; 1983.

% of current expenditure total	% of capital expenditure in total	Total federal budget deficit
34	56	−1,463.4
75	25	−458.9
33	77	−1,947.2
63	37	96.7
57	43	−2,743.6
56	44	−2,935.6
53	43	−3,652.9
57	43	−4,867.7
63	37	−2,278.8
64	36	−8,461.1

Table B.8 Gross Domestic Product (millions of naira)

	Total GDP (incl. oil)			Total GDP (excl. oil)		
	Current	Constant	GDP deflator*	Current	Constant	GDP deflator*
1973	11,223.6	11,223.6		9,400.5	9,400.5	
1974	18,652.0 (66.2)†	12,194.5 (8.7)	57.5	12,813.1 (36.3)	10,156.9 (8.0)	28.3
1975	21,275.1 (15.1)	12,500.5 (2.5)	12.6	17,196.7 (34.2)	10,924.2 (7.6)	26.6
1976	27,317.8 (27.2)	13,744.3 (10.0)	17.2	21,211.9 (23.3)	11,871.2 (8.7)	14.6
1977	32,051.8 (17.3)	14,749.2 (7.3)	10.0	24,980.2 (17.8)	14,730.2 (24.1)	−6.3
1978	33,660.4 (5.0)	13,966.9 (−5.3)	10.3	26,092.0 (4.5)	13,949.6 (−5.3)	9.8
1979	39,938.6 (18.7)	14,618.4 (4.7)	14.0	29,459.4 (12.9)	12,602.1 (−9.7)	22.6
1980	43,280.2 (8.4)	14,874.3 (1.8)	6.6	32,545.9 (10.5)	13,018.4 (3.3)	13.8
1981	43,450.0 (0.4)	14,571.5 (−2.0)	2.4	35,348.3 (8.6)	13,304.8 (2.2)	6.4

Source: Federal Office of Statistics, *Nigerian Gross Domestic Product and Allied Macro–Aggregates*, 1982.

*The GDP deflator has been derived by subtracting the percentage change in current prices from the percentage change in constant prices.

†Figures in parentheses refer to percentage changes.

Table B.9 Components of GDP (millions of naira)

	Agriculture			Manufacturing and mines (excl. oil)			Construction			Services		
	Current	Constant	GDP deflator*	Current	Constant	GDP deflator*	Current	Constant	GDP deflator*	Current	Constant	GDP deflator*
1973	3,371.5	3,371.5		197.5	197.5		1,123.2	1,123.2		4,166.3	4,166.3	
1974	4,941.9†	3,718.4	36.3	248.3	209.2	19.8	1,315.7	1,131.7	16.3	5,592.0	4,571.3	24.5
	(46.6)	(10.3)		(25.8)	(6.0)		(17.1)	(0.8)		(34.2)	(9.7)	
1975	5,872.8	3,340.0	29.0	390.0	326.4	1.0	1,814.6	1,129.7	38.1	7,885.5	5,586.4	18.8
	(18.8)	(−10.2)		(57.0)	(56.0)		(37.9)	(−0.2)		(41.0)	(22.2)	
1976	6,783.3	3,307.1	16.5	691.4	406.5	52.7	2,605.8	1,497.0	11.1	9,591.1	5,878.6	16.4
	(15.5)	(−1.0)		(77.3)	(24.6)		(43.6)	(32.5)		(21.6)	(5.2)	
1977	8,073.7	3,502.9	13.1	833.4	473.0	4.2	2,990.8	1,748.4	−2.0	11,428.7	6,291.4	12.2
	(19.0)	(5.9)		(20.5)	(16.3)		(14.8)	(16.8)		(19.2)	(7.0)	
1978	8,339.0	3,218.7	11.4	847.2	449.2	6.7	3,077.2	1,681.2	6.7	11,915.8	5,928.2	10.1
	(3.3)	(−8.1)		(1.7)	(−5.0)		(2.9)	(−3.8)		(4.3)	(−5.8)	
1979	8,980.9	3,135.1	10.3	860.3	430.4	5.7	3,192.3	1,624.5	7.1	14,227.6	6,358.3	12.1
	(7.7)	(−2.6)		(1.5)	(−4.2)		(3.7)	(−3.4)		(19.4)	(7.3)	
1980	9,619.6	3,067.7	9.3	879.8	422.0	4.2	3,671.2	1786.5	5.0	15,817.7	6,519.4	8.7
	(7.1)	(−2.1)		(2.3)	(−1.9)		(15.0)	(10.0)		(11.2)	(2.5)	
1981	10,137.7	2,982.5	8.2	895.6	411.2	4.4	4,001	1,873.1	4.1	17,404.4	6,673.7	7.6
	(5.4)	(−2.8)		(1.8)	(−2.6)		(9.0)	(4.9)		(10.0)	(2.4)	

Source: As table B.8.
†Figures in parentheses refer to percentage changes.
*The GDP deflator has been derived by subtracting the percentage change in current prices from the percentage change in constant prices.

Table B.10 Gross Domestic Product at 1973–4 factor cost, percentage distribution

	Activity sectors	1973–4	1974–5	1975–6	1976–7	1977–8	1978–9	1979–80	1980	1981
1.	Agriculture	19.84	20.68	17.60	15.16	15.39	14.10	13.02	12.24	11.83
2.	Livestock	3.54	3.33	3.12	3.14	3.10	3.38	3.03	3.02	3.13
3.	Forestry	2.42	2.43	2.16	2.00	1.70	1.69	1.59	1.54	1.53
4.	Fishing	4.23	4.05	3.83	3.76	3.56	3.87	3.81	3.83	3.98
5.	Mining and quarrying	18.00	18.43	14.42	16.59	16.07	15.61	16.74	15.31	11.52
5.1	Coal	0.01	0.00	0.00	0.00	0.00	0.00	0.00	0.00	0.00
5.2	Crude petroleum	16.21	16.71	12.61	13.63	12.87	12.39	13.79	12.48	8.69
5.3	Metal ore	0.16	0.13	0.10	0.07	0.06	0.07	0.06	0.05	0.06
5.4	Quarrying and metallic minerals	1.59	1.59	1.71	2.89	3.14	3.15	2.88	2.78	2.77
6.	Manufacturing	4.43	3.94	4.75	5.33	5.28	6.37	6.53	7.55	8.61
6.1	Large scale	3.95	3.52	4.24	4.76	4.71	5.69	5.83	6.75	7.70
6.2	Small scale	0.48	0.42	0.51	0.57	0.57	0.68	0.70	0.80	0.91
7.	Utilities	0.40	0.37	0.40	0.36	0.39	0.49	0.54	0.62	0.75
7.1	Electricity	0.39	0.35	0.37	0.33	0.36	0.45	0.48	0.55	0.66
7.2	Water	0.01	0.02	0.03	0.03	0.03	0.04	0.06	0.07	0.09
8.	Building and construction	10.01	9.28	9.04	10.89	11.85	12.04	11.11	12.01	12.85
9.	Transport	3.73	3.43	3.45	3.27	3.15	3.54	3.63	3.98	4.54
9.1	Road	3.35	3.09	3.02	2.79	2.68	2.88	2.85	3.02	3.33
9.2	Rail	0.13	0.06	0.08	0.06	0.06	0.07	0.08	0.08	0.10
9.3	Ocean	0.19	3.19	0.21	0.28	0.27	0.42	0.51	0.66	0.85
9.4	Air	0.05	0.09	0.14	0.14	0.14	0.17	0.19	0.22	0.26
10	Communication	0.24	0.24	0.30	0.30	0.29	0.34	0.35	0.35	0.38
10.1	Posts, telephone and telegraphs	0.21	0.19	0.22	0.23	0.22	0.25	0.26	0.26	0.28
10.2	Radio broadcasting, television	0.03	0.05	0.08	0.08	0.07	0.09	0.09	0.09	0.10
11.	Wholesale and retail trade	19.58	19.20	21.38	20.85	21.45	20.75	22.09	21.77	22.19
12.	Hotels and restaurants	0.29	0.28	0.29	0.28	0.28	0.36	0.33	0.41	0.46
13.	Finance and insurance	1.25	2.20	3.08	2.98	3.14	3.57	3.54	3.62	3.83
13.1	Financial institutions	1.09	2.03	2.83	2.66	2.78	3.05	2.87	2.89	3.02
13.2	Insurance	0.16	0.17	0.25	0.32	0.36	0.52	0.67	0.73	0.82
14.	Real estate and business services	0.54	0.54	0.62	0.52	0.51	0.56	0.54	0.54	0.57
14.1	Real estate	0.02	0.04	0.04	0.03	0.03	0.04	0.03	0.03	0.02
14.2	Professional services	0.52	0.50	0.58	0.48	0.48	0.52	0.51	0.51	0.55
15.	Housing	5.58	5.98	6.11	5.65	5.35	5.61	5.38	5.35	5.54
16.	Sub-total	94.08	94.38	90.55	91.08	91.51	92.28	92.28	92.15	91.71
17.	Producers of government services	5.92	5.62	9.45	8.92	8.49	7.72	7.72	7.86	8.29
17.1	Government	5.41	5.12	8.56	7.96	7.54	6.58	6.45	6.82	6.82
17.2	University	0.44	0.37	0.69	0.68	0.63	0.69	0.67	0.67	0.73
17.3	Others	0.07	0.13	0.20	0.28	0.32	0.45	0.58	0.65	0.74
18.	Total	100.0	100.0	100.0	100.0	100.0	100.0	100.0	100.0	100.0

Source: As Table B.8.

Table B.11 Indices of wages in different activities

	Nominal					Real*				
	Agri-culture	Manu-facturing	Const-ruction	Govern-ment service	Minimum wage	Agri-culture	Manu-facturing	Const-ruction	Govern-ment service	Minimum wage
1970					76.6					96.6
1971					94.2					102.3
1972	98.9	91.5	100.0		100.0	104.7	96.8	105.8		105.8
1973	100.0	100.0	100.0	100.0	100.0	100.0	100.0	100.0	100.0	100.0
1974	123.6	109.9	200.0	135.2	100.0	110.0	97.8	177.9	120.3	89.0
1975	177.5	124.6	200.0	203.6	200.0	118.1	82.9	133.1	135.5	133.1
1976		138.0	200.0	224.7	200.0		73.8	107.0	120.2	107.0
1977		152.1	213.0	252.3	214.0		68.2	95.5	113.1	96.0
1978	268.5	162.0	300.0	234.6		101.5	61.2	113.4	88.7	
1979	295.5	232.4	330.0	263.0		100.5	79.0	112.2	89.4	
1980	522.5	302.8	347.0	291.7		159.5	92.4	105.9	89.0	

Sources: Nominal wage indices calculated from ILO, *Statistical Yearbook*, 1982, except government service index, calculated from Federal Office of Statistics, *Nigerian GDP and Allied Macro-Aggregates*, 1982, and minimum wage index calculated from G. Williams,[6] 'State and Society in Nigeria'[12], Afrografika, 1978, p.164.

*Real-wage indices were calculated by dividing indices of nominal wage by CPI.

Table B.12 Index of manufacturing production (weights: 1972 value added)

	Weights	1971	1972	1973	1974	1975	1976	1977	1978	1979*	1980†	Growth rate 1971–80 (%)
Vegetable oil	6.2	118.5	100.0	151.2	43.8	35.7	24.4	4.8	15.4	16.3	12.5	−22.1
Sugar	2.6	112.4	100.0	98.6	106.5	125.8	88.7	123.2	111.8	116.2	122.2	0.9
Sugar confectionery	4.5	61.2	100.0	101.0	86.9	118.4	127.1	207.3	201.6	199.4	210.7	14.7
Soft drinks	5.6	76.7	100.0	161.5	154.0	224.9	322.1	303.5	332.2	433.5	486.3	22.8
Beer (including stout)	28.2	84.6	100.0	130.5	148.4	178.5	191.0	185.6	285.2	310.8	364.3	17.6
Cigarettes	20.3	95.3	100.0	89.7	96.7	107.0	128.3	122.0	129.0	117.3	116.8	2.3
Cotton textiles	24.4	120.3	100.0	127.3	118.8	144.9	161.0	172.9	167.1	184.2	196.2	5.6
Other textiles	4.5	45.3	100.0	133.5	393.7	611.0	1051.8	964.7	1129.3	1297.1	1412.9	46.6
Footwear	1.6	109.9	100.0	85.0	111.4	122.6	110.0	123.5	119.3	128.8	127.9	1.7
Paint and allied products	1.8	92.0	100.0	122.2	113.8	151.7	180.2	241.8	280.3	274.6	316.0	14.7
Soap and detergent	12.6	88.5	100.0	161.3	168.6	177.9	228.1	328.4	362.5	325.5	340.3	16.1
Refined petroleum products	4.8	45.5	100.0	127.0	124.4	105.4	128.0	123.6	124.5	150.6	189.3	17.2
Other petroleum products	17.0	95.0	100.0	118.7	40.6	160.8	84.9	86.6	74.6	93.1	119.9	2.6
Pharmaceuticals	1.8	109.7	100.0	141.2	84.9	148.3	239.8	186.5	352.9	227.2	238.7	9.0
Rubber	1.1	117.2	100.0	78.5	68.3	140.1	99.6	109.3	122.6	112.6	113.8	−0.3
Cement	6.0	65.8	100.0	112.5	108.7	115.6	115.4	117.1	139.6	161.9	162.0	10.5
Tin metal	1.4	105.1	100.0	85.5	79.2	161.4	54.3	47.4	42.6	40.9	40.0	−10.2
Roofing sheets	3.5	90.9	100.0	114.2	82.3	137.9	161.2	214.7	191.7	218.6	250.0	11.9
Vehicle assembly	0.7	120.3	100.0	118.4	130.7	302.2	698.6	1097.3	992.7	1138.9	1363.9	31.0
Radio, changers, TV assembly	1.3	135.3	100.0	97.9	66.2	108.0	119.1	128.6	95.7	187.4	240.8	6.6
Total	150.2	92.8	100.0	123.6	119.5	147.7	182.2	193.5	221.4	237.5	263.6	12.3

*Revised
† Estimated
Source: Central Bank of Nigeria, Annual Report 1977; 1979; 1980.

Table B.13 Index of industrial production

	1970	1971	1972	1973	1974	1975	1976	1977	1978	1979	1980	Compound growth, 1973–80 (%)
Mining and quarrying	42.9	60.5	71.4	81.4	100.7	79.4	97.3	100.0	91.6	104.6	96.8	2.2
Manufacturing (large-scale)	46.8	45.0	57.4	63.8	61.7	76.3	94.1	100.0	114.3	122.7	144.6	10.8
Manufacturing (small-scale)	62.1	62.7	63.2	63.8	61.7	76.3	94.1	100.0	114.3	122.7	142.1	10.5
Building and construction	29.4	41.5	51.2	64.2	64.7	64.6	85.6	100.0	96.1	92.9	102.1	6.0
Electricity	44.3	56.1	68.1	81.8	81.7	87.6	85.9	100.0	116.5	132.8	152.7	8.1
Overall index	40.4	54.2	64.9	75.1	87.2	75.5	94.0	100.0	95.6	104.2	104.4	4.2

Source: As Table B.8.

Table B.14 Nigeria – price level, food prices and real exchange rate (1969 – 1981)

	IMF–Nigeria Consumer price index	Food–Price CPI Price Relative	World Prices US Index	Real appreciation exchange rate*
1969	69.7	93.1	82.4	77.9
1970	79.3	101.1	87.3	83.7
1971	92.1	109.1	91.0	93.5
1972	94.5	108.0	94.1	100.5
1973	100.0	100.0	100.0	100.0
1974	112.4	106.8	110.9	106.0
1975	150.3	114.9	121.1	132.7
1976	186.9	114.9	128.1	153.2
1977	223.0	128.7	136.4	166.7
1978	264.6	127.5	146.7	186.7
1979	294.1	122.9	163.3	196.1
1980	327.6	122.9	185.4	212.7
1981	395.7	132.1	204.6	206.0

Sources: IMF Nigeria Index (1984); food-price index, calculated from ILO Yearbook; World index (US price index) from, World Bank Country Tables.

* The real appreciation exchange rate is calculated as

$$\frac{\text{consumer price index}}{\text{U.S. price index}} \times \text{exchange rate (unit/\$)}$$

Table B.15 Agricultural production by commodity (kilotonnes)

	1971	1972	1973	1974	1975	1976	1977	1978	1979*	1980†	Average compound growth(%)
Rice, paddy	462	466	514	523	600	611	620	826	870	1,030	8.3
Corn	1,042	1,182	1,287	1,350	1,400	1,440	1,500	1,640	1,670	1,720	5.1
Millet	2,688	3,048	2,150	2,800	2,865	2,865	2,950	3,100	3,130	3,200	1.8
Sorghum	3,140	3,561	2,968	3,500	3,590	3,680	3,750	3,760	3,785	3,800	1.9
Pulses	542	540	480	525	540	555	365	450	480	490	-1.0
Cassava	12,396	12,700	13,000	13,300	13,600	13,900	14,000	14,150	14,600	14,800	1.8
Yams	16,104	16,257	16,800	17,200	17,600	18,000	18,000	18,100	18,100	18,100	1.2
Cocoyams	1,479	1,524	1,565	1,600	1,640	1,680	1,700	1,710	1,710	1,710	1.5
Tobacco	15	13	12	12	18	10	8	8	12	17	1.3
Cotton	38	49	30	52	58	81	36	37	29	40	0.5
Cottonseed	77	95	64	106	106	130	70	80	52	69	-1.0
Soybeans	1	4	1	1	1	1	3	3	3	3	11.6
Peanuts, in shell	845	1,125	340	530	332	350	643	469	400	400	-7.2
Sesame seeds	3	7	3	6	6	6	6	6	6	6	7.2
Bananas and plantains	1,300	1,330	1,360	1,390	1,420	1,450	1,400	1,425	1,425	1,425	0.9
Other fruit	47	48	50	51	52	53	53	65	65	65	3.3
Coffee	4	4	2	2	4	3	3	3	3	3	-2.8
Cocoa beans	265	264	218	213	218	167	205	140	170	170	-4.3
Rubber	62	57	66	78	68	56	60	65	65	66	0.6
Kola nuts	136	139	143	146	150	154	152	160	160	160	1.6
sugar, raw	34	40	40	60	50	40	36	34	29	32	-0.6
Palm oil	432	457	432	491	500	500	510	515	500	520	1.9

Table B.15 Cont'd

	1971	1972	1973	1974	1975	1976	1977	1978	1979*	1980†	Average compound growth(%)
Palm kernels	307	295	244	310	295	321	340	350	335	345	1.2
Meats	555	555	565	555	575	590	545	550	550	550	-0.1
Milk	381	381	360	355	360	370	370	370	380	380	0.0

Source: US Department of Agriculture, Economic Research Services, cited in USDA (1981, p. 19).

Table B.16 Guaranteed minimum food prices (naira per tonne)

	1970	1971	1972	1973	1974	1975	1976	1977	1978	1979	1980	1981	Compound growth (%)
Soybeans	35	35	47	49	99			130	135	150		321	15.7
Millet	42*						80	110	110	220	220		15.6
Beans								180	180	345	345	362	15.0
Sorghum	35*						80	110	110	210	210	220	14.0
Cassava	42	77	66	53			85	110	110				11.3
Maize	68	99	93	90			95	130	130	200	200	210	9.9
Rice, milled								400	400	570	570	596	8.3
Rice, paddy	136	174	176	191			185	240	240	329	329	345	8.1
Yams	64	95	115	126			85	120	120				7.2
Wheat							174F	174F	200	235	235	247	6.0

Sources: 1976–80 data from World Bank (1981, p.117); other data from FAO ICS data base, Rome.
*1968

235

Table B.17 Supply and distribution of rice

	Area harvested thousands of hectares	Yield (tonnes/ha)	Rough production thousands of tonnes
1970	254	1.68	427
1971	263	1.76	462
1972	275	1.70	466
1973	280	1.84	514
1974	285	1.84	523
1975	300	1.95	586
1976	310	1.97	611
1977	325	1.99	647
1978	414	2.00	827
1979	428	2.11	902
1980	550	2.05	1,090

Source: USDA (1981).

Table B.18 Supply and distribution of maize

	Area Harvested (thousands of hectares)	Yield (tonnes/ha) tonnes	Beginning stocks (thousands of tonnes)	Production (thousands of tonnes)	Total imports (thousands of tonnes)	Total exports	Domestic consumption (thousands of tonnes)	
							For feed	Total
1970–71	1,260	1.04	–*	1,310	10	–	30	1,320
1971–72	1,270	0.73	–	931	2	–	18	933
1972–73	1,400	0.84	–	1,182	2	–	22	1,184
1973–74	1,560	0.83	–	1,287	2	–	25	1,289
1974–75	1,625	0.83	–	1,350	3	–	25	1,353
1975–76	1,675	0.84	–	1,400	1	–	30	1,401
1976–77	1,725	0.83	–	1,440	25	–	30	1,465
1977–78	1,800	0.83	–	1,500	75	–	70	1,530
1978–79	1,820	0.90	45	1,640	40	–	120	1,660
1979–80	1,850	0.90	65	1,670	125	–	185	1,785
1980–81	1,900	0.89	75	1,720	160	–	250	1,885

Source: USDA (1981)
*Denotes negligible or zero.

Table B.19a Millet

	Area harvested (thousands of hectares)	Yield (hg/ha)	Production (thousands of tonnes)	Imports thousands of tonnes)	1000 naira
1970	4,926	5609	2,763	58	5,200
1971	5,275	5585	2,946	58	5,300
1972	4,839	6299	3,048	50	5,400
1973	4,700	5000	2,350	30	3,300
1974	4,800	5333	2,800	56	5,600
1975	4,800	5969	2,865	30	4,000
1976	4,800	5969	2,865	40	6,000
1977	4,920	5996	2,950	40	6,500
1978	5,000	6200	3,100	30	6,000
1979	5,000	6260	3,130	20	5,500
1980	5,030	6223	3,130	20	5,500
1981	5,050	6396	3,230	30	6,500

Source: FAO/ICS data base, Rome.

Table B.19b Sorghum

	Area harvested (thousands of hectares)	Yield (hg/ha)	Production (thousands of tonnes)	Producer (naira per kilo)
1970	5,670	6730	3,816	52
1971	5,409	5805	3,140	76
1972	5,472	6508	3,561	71
1973	5,300	6226	3,300	69
1974	5,645	6200	3,500	83
1975	5,795	6195	3,590	95
1976	5,940	6195	3,680	110
1977	6,000	6250	3,750	110
1978	6,000	6267	3,760	110
1979	6,000	6308	3,785	126
1980	6,000	6333	3,800	210
1981	6,000	6250	3,750	0

Source: FAO/ICS data base, Rome.

Table B.20 Exports of major non-oil commodities by economic sectors (kilotonnes)

	1970	1971	1972	1973	1974	1975	1976	1977	1978	1979	1980
Major agricultural products											
Cocoa	195.7	271.7	227.5	213.8	194.0	174.7	218.9	167.5	191.7	217.8	157.1
Cotton (raw)	28.2	22.3	1.0	8.3	–	–	–	9.2	3.2	2.6	2.4
Groundnuts	292.3	137.5	106.1	198.7	30.3	–	1.6	0.8	–	–	–
Groundnut oil	90.0	41.3	39.7	110.8	23.5	0	–	–	–	–	–
Hides and skins	4.7	3.8	4.3	5.4	5.3	2.9	2.1	2.0	1.1	0.7	0.6
Palm oil	185.3	241.1	212.2	137.6	0	10.7	3.3	–	3.2	0.7	–
Palm kernel	7.6	20.2	1.9	0	185.6	171.4	272.0	186.0	56.8	50.8	49.6
Rubber (natural)	61.7	51.3	41.2	49.4	61.3	60.9	34.0	27.7	30.9	34.2	31.0
Timber (log and sawn)	221.5	207.1	231.9	25.6	289.7	105.3	28.7	11.9	0	0	–
Coffee	1.7	3.7	4.2	2.5	0.3	1.1	6.0	2.0	0.7	–	–
Mineral products											
Columbite	1.7	1.2	4.6	2.1	2.3	1.0	1.1	1.2	0.6	0.6	0.6
Others	–	–	–	–	21.4	2.9	2.5	4.0	3.6	3.9	3.9
Manufactures and semi-manufactures											
Agricultural	179.1	114.7	117.7	164.5	54.6	26.4	34.6	26.2	9.7	10.8	9.9
Cocoa butter	9.4	8.3	10.3	11.1	11.1	9.3	5.9	7.7	4.2	5.0	4.6
Cocoa powder	0.5	0.6	–	0.5	1.6	0.5	1.4	1.4	1.8	2.2	1.6
Cocoa cake	9.0	8.3	8.1	15.4	11.4	9.7	4.5	6.8	3.7	3.6	3.0
Groundnut cake	160.2	97.5	99.3	137.5	30.5	6.9	28.7	8.4	–	–	–
Tin metal	10.9	8.1	6.8	5.2	5.8	4.7	3.4	1.7	1.3	1.5	1.4

Source: Central Bank of Nigeria, Annual Report 1972, p. 86; 1973, p. 92; 1974, p. 86; 1976, p. 75; 1977, p. 77; 1978, p. 75; 1979, p. 73; 1980, p. 89; 1981, p. 86; 1982, p. 85.

Table B.21 World price index of major commodities

	1970	1971	1972	1973	1974	1975	1976	1977	1978	1979	1980
Major agricultural projects											
Cocoa	25.9	20.7	24.7	43.4	59.9	47.9	78.6	145.6	130.8	126.5	100.0
Cotton (raw)	30.9	34.1	42.2	69.0	71.2	55.5	83.6	75.8	70.8	76.4	100.0
Groundnuts	47.0	51.7	52.3	80.6	152.2	89.2	87.1	112.6	129.9	115.9	100.0
Hides and skins	28.1	31.6	64.5	74.7	51.4	50.8	73.2	80.6	102.8	159.5	100.0
Palm oil	44.6	44.8	37.3	64.7	114.7	73.6	69.7	92.3	103.0	112.1	100.0
Palm kernel	48.6	42.1	33.7	75.1	134.8	60.0	66.7	94.7	105.6	145.0	100.0
Rubber (natural)	28.7	24.5	24.6	47.9	54.1	40.7	53.9	56.6	69.0	87.3	100.0
Timber (log and sawn)	23.0	23.1	21.9	36.5	43.5	36.0	49.0	47.9	48.9	85.4	100.0
Coffee	33.5	29.6	33.4	41.2	45.1	48.1	94.2	152.0	102.8	112.5	100.0
Manufactures and semi-manufactures											
Groundnut cake	42.3	40.7	50.6	110.4	72.2	58.1	73.0	88.8	85.1	87.6	100.0
Tin metal	21.9	20.8	22.4	28.7	48.7	40.7	45.4	64.1	76.3	91.8	100.0

Source: IMF (1983, pp. 93–5).

Table B.22 Estimated palm oil production (kilotonnes)

	Domestic edible oil	Market board purchases*	Total oil	Total kernel
1960	355	190	545	423
1961	365	173	538	430
1962	376	129	505	362
1963	387	149	536	413
1964	398	148	546	401
1965	410	164	574	449
1966	422	158	580	422
1967	288	32	320	223
1968	322	4	326	194
1969	395	23	418	172
1970	463	25	488	265
1971	489	31	520	307
1972	490	20	510	269
1973	411	14	425	230
1974	457	26	483	302
1975	448	52	500	274
1976	460	55	515	147
1977	463	47	510	186
1978	515	–	515	325
1979	500	–	500	350
1980†	520	–	520	–

Source: Kilby (1981).
*Exports and domestic sales to margarine and soap manufacturers.
†Projected.

Table B.23 Palm produce prices (naira per tonne)

	World prices		Producer prices*	
	Oil	Kernel	Oil	Kernel
1960			106	58
1961			106	58
1962			80	50
1963	142	98	80	50
1964	154	97	82	54
1965	182	119	82	54
1966	152	111	86	54
1967	149	95	86	54
1968	94	113	84	54
1969	101	99	84	54
1970	167	106	82	58
1971	172	95	90	60
1972	143	76	92	61
1973	249	170	100	130
1974	413	287	220	133
1975	271	129	280	150
1976	256	145	280	150
1977	345	212	355	150
1978	389	236	355	150
1979	374	298	355	150
1980	–	–	355	180

Source: Kilby (1981).
*The statutory producer price for palm oil has not been the effective domestic price, which has been much higher, since 1967. Such purchases as the market board has made have been from government plantations.

Table B.24 Groundnut

	Production (kilotonnes)	Area harvested (thousands of hectares)	Yield (tonnes/hectare)	Exports (kilotonnes)	Producer price (naira/tonne)
1966	1,693	2,254	0.75	582	60
1967	1,558	2,256	0.69	549	51
1968	1,813	1,941	0.93	648	34
1969	1,846	1,833	1.00	525	41
1970	1,581	1,870	0.84	291	47
1971	1,554	1,870	0.84	136	55
1972	945	1,834	0.65	196	55
1973	350	1,429	0.36	198	65
1974	400	970	0.41	303	175
1975	280	970	0.30	0	175
1976	500	920	0.54	16	175
1977	300	920	0.36	8	275
1978	450	820	0.75	0	290
1979	540	600	0.90	19	350
1980	570	600	0.95	0	420

Source: FAO/ICS data base, Rome.

Table B.25 Production of major food crops (thousands of tonnes)

	Yams			Cassava			Millet			Sorghum			Maize			Rice			Wheat		
	USDA	FAO	FOS	USDA	FAO	FOS	USDA	FAO	FOS	USDA	FAO	FOS	USDA	FAO	FOS	USDA	FAO	FOS	USDA	FAO	FOS
1961							2,600	2,600		3,966	3,966		1,000	1,000		354	119			16	
1962							2,530	2,530		4,509	4,509		900	900		370	258			16	
1963							2,732	2,732		4,069	4,069		1,050	1,050		304	196			16	
1964							2,484	2,484		4,239	4,239		1,090	1,090		405	221			13	
1965							2,729	2,729		4,325	4,325		1,040	1,040		355	232			13	
1966							1,747	1,747		3,160	3,160		1,020	1,020		406	200			13	
1967							2,590	2,590		3,389	3,389		1,000	1,000		391	386			10	
1968							2,196	2,196		2,821	2,821		950	950		375	353		5	10	
1969							3,298	3,298		4,080	4,080		1,426	1,426		386	235		6	10	
1970	15,200	NA	8,230	11,871	9,084	5,180	3,284	3,284	3,077	4,080	4,080	4,044	1,310	1,310	1,376	427	490	293	6	7	19
1971	16,104		7,990	12,396	9,172	4,719	2,688	2,688	2,911	3,140	3,140	3,265	931	931	1,322	462	580	279	7	7	20
1972	16,257		7,758	12,700	9,570	3,156	3,048	3,414	2,524	3,561	3,988	3,367	1,182	1,188	830	466	600	397	7	7	20
1973	16,800		6,911	13,000	9,600	2,675	2,150	2,350	2,882	2,968	2,968	2,546	1,287	441	690	514	343	459	4	4	15
1974	17,200		7,003	13,300	10,000	3,115	2,800	3,000	4,322	3,500	3,500	3,609	1,350	980	724	523	356	498	6	6	18
1975	17,600		7,598	13,600	10,500	3,223	2,865	3,000	4,653	3,590	4,204	1,400	1,260	769	586	515	522	7	18	18	
1976	18,000		7,948	13,900	10,800	3,025	2,865	2,900	2,653	3,680	3,680	2,933	1,440	1,295	1,253	611	534	426	7	20	20
1977	18,000		6,342	14,000	10,600	1,759	2,950	2,950	2,779	3,750	3,750	3,051	1,550	1,350	943	620	647	276	8	21	21
1978	18,100		6,223	14,150	10,500	1,674	3,100	3,060	2,521	3,770	3,770	3,023	1,640	1,480	643	826	842	372	8	21	22
1979	18,100		5,588	14,600	10,500	1,581	3,130	3,130	2,171	3,785	3,785	2,719	1,670	1,500	608	871	750	235	10	21	22
1980	18,180		5,120	13,100	11,000	1,590	3,130	3,130	2,756	3,800	3,800	3,303	1,720	1,550	596	1,090	1,090	245	15	24	24
1981	18,200		5,120	11,800	11,000	1,700	3,180	3,230	2,906	3,750	3,750	3,475	1,750	1,580	602	1,165	1,241	257	20	25	26
1982	18,480		5,397	11,700	11,500	1,775	3,240	3,300	3,034	3,825	3,800	3,626	1,775	1,650	628	1,380	1,400	288	25	25	27

USDA: Reports P82 (18 January 1983). P8 (30 January 1978) plus telephone updates for root crops. Table 135 – Nigeria; for yam 1980-82, report by R. D. Norton, 'Pricing Policy Analysis for Nigerian Agriculture'. Sept. 1983, p. 2.2.
FAO: *Production Yearbook*, 1975, ff.
FOS: Central Bank of Nigeria, worksheets, prepared for the Mission in May 1981.
Source: World Bank, Agricultural Sector Memorandum, Main Report, Vol. II, June 1984.

Table B.26 Average producer prices of major export commodities (naira per tonne)

Crop	1972	1973	1974	1976	1978	1980	1982	Compound nominal growth per year (%)	Compound real growth per year (%)
Benniseed	81	105	169	264	300	315	315	14.5	−2.0
Cocoa	297	354	487	660	1,030	1,300	1,300	15.5	−1.0
Cotton (seed)	123	132	156	308	330	400	510	15.5	−1.0
Groundnuts	75	80	145	250	290	420	450	19.5	+3.0
Palm kernels	61	61	124	150	150	180	230	14.0	−2.5
Palm oil	76	84	204	265	355	450	495	20.5	+4.5
Soy beans	37	47	60	99	175	150	175	16.5	0
Coffee (Arabica)	–	–	–	700	1,100	1,155	1,155	8.5	−10.0
Rubber	–	–	–	–	575	795	1,200	30	+25

Sources: Central Bank of Nigeria, *Annual Report*, 1972, p. 23; 1974, p. 22; 1976, p. 12; 1978, p. 11; 1982, p.10; Final column based on deflating nominal growth of prices by the Consumer Price Index indicated in IMF (1984).

Table B.27 Cocoa beans

	Area harvested (thousands of hectares)	Yield (hg/ha)	Production (tonnes)	Producer price naira/T	Real producer price index	Exports (tonnes)	1000 Naira
1968	600	3,197	191,800	179		208,882	144,874
1969	600	3,680	220,800	189		173,605	147,269
1970	750	4064	304,800	288	91.2	195,907	186,305
1971	700	3664	256,600	297	122.1	271,738	201,790
1972	700	3444	241,100	297	108.4	227,532	153,724
1973	720	2986	215,000	297	105.7	213,897	170,796
1974	720	2972	214,000	541	100.0	197,125	252,444
1975	720	2986	215,000	660	161.9	194,692	294,050
1976	720	2292	165,000	660	147.7	222,966	349,327
1977	720	2806	202,000	1,030	118.8	167,521	482,600
1978	700	2143	150,000	1,030	155.5	192,761	638,328
1979	700	2286	160,000	1,200	131.0	257,112	436,886
1980	700	2214	155,000	1,300	137.3	133,861	412,000

Source: FA/ICS data base, Rome.

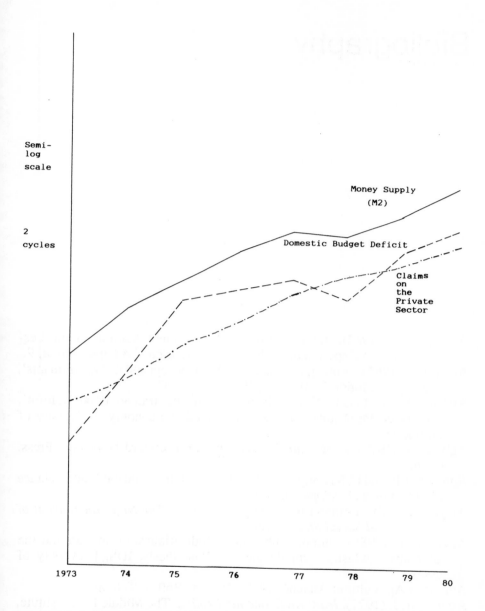

Figure B.1 Money supply and its determinants, Nigeria, 1973–80

Bibliography

Abalu, G.O.I. (1982), 'Integrated Rural Development Administration, Lessons from Two Experiments in Nigeria', *Agricultural Administration, 9.*
Adler, J.H. (1965), 'Absorptive Capacity: The Concept and its Determinants', Brooking Institution Staff Paper, Washington DC.
Afshar, Haleh (1972), 'Comparison of Irish and Iranian Land Reform', unpublished Ph.D. thesis Department of Land Economy, University of Cambridge.
Agbola, A. (1979), *Agricultural Atlas of Nigeria,* Oxford University Press, Oxford.
Aghevli, B.B. and C. Sassanpur (1982), 'Prices, Output, and the Trade Balance in Iran', *World Development,* September.
Ajayi, I. (1978), 'Demand for money in Nigeria', *The Nigerian Journal of Economic and Social Studies,* July.
Alizadeh, P. (1985), 'Import Substitution Industrialisation in Iran and the Automative Industry', unpublished D. Phil. thesis, IDS, University of Sussex.
American Agricultural Attaché (1977), Report IR6031, Tehran.
Amuzegar, J. (1977), *Iran: An Economic Profile.* The Middle East Institute, Washington DC.
Amuzegar, J. (1982), 'Oil Wealth: A Very Mixed Blessing', *Foreign Affairs,* Spring, 814–35.
Arief, S. and J. Sudaram (1983), *The Malaysian Economy and Finance.* Rosecons, Australia.
Ashraf, A. and A. Banuazizi (1980), 'Policies and Strategies of Land Reform

in Iran' in 'Inayatullah, ed., *Land Reform: Some Asian Experiences*. APDAC, Kuala Lumpur.

Avramovic, Dragoslav 1970, 'Industrialization in Iran: the Records, the Problems, and the Prospects', *Tahgigate Egtesadi* (Quarterly Journal of Economic Research), Spring.

Awa, N.E. (1971), *Food Production Problems of Small Farmers in Low-Technology Nations: Some Evidence from Nigeria.* Dept. of Communication Arts, Cornell University, New York.

Awojobi, A. (1982), 'Where Our Oil Money Has Gone', published lecture delivered in 1982, University of Ife, Lagos.

Bhagwati, J. (1984), "Why Services are Cheaper in Poor Countries', *Economic Journal,* March.

Booth, A. and McCawley, P. (1981), *The Indonesian Economy During the Soeharto Era.* Oxford University Press, Kuala Lumpur.

Carter, R.C. *et al.* (1983), 'Policies and Prospects in Nigerian Irrigation', *Outlook on Agriculture, 12,* 2.

Caves R. and Jones, R. (1983), *World Trade and Payments,* Little Brown, Boston, MA, 3rd Edn.

Center for Agricultural Marketing Development (CAMD) (1977), *A Study of the Rice Market,* Tehran.

Central Bank of Nigeria (1970-9), *Annual Report and Balance Sheets.*

Central Bank of Nigeria (1983a), *Annual Report and Balance Sheet.*

Central Bank of Nigeria (1983b), *'Nigeria's Flow of Funds, 1970–78.*

Chavanian, M.M.H. (1975), 'New Approaches to Agricultural Development and Investment Opportunities in Iran', address to International Agricultural and Food Exhibition, Tehran.

Collier, P. (1983), 'Oil and Inequality in Rural Nigeria' in D. Ghai and S. Radwan, eds, *Agrarian Policies and Rural Poverty in Africa.* ILO, Geneva.

Corden, W.M. (1966), 'The Structure of a Tariff System and the Effective Protection Rate', *Journal of Political Economy, 74,* June 221–37.

Corden, W.M. (1974), *Trade Policy and Economic Welfare.* Oxford University Press, London.

Corden, W.M. (1982), 'Booming Sector and Dutch-Disease Economics: A Survey, Australian National University, Working Papers in Economics, 79, November.

Corden, W.M. and J.P. Neary (1982), 'Booming Sector and De-industrialisation in a Small Open Economy'. *Economic Journal,* December.

Corden, W.M. and P.G. Warr (1981), 'The Petroleum Boom and Exchange-Rate Policy in Indonesia: A Theorectical Analysis', *Ekonomi Dakeuangan Indonesia, 29,* 3, September.

Crockett, A. and O. Evans (1980), 'Demand for Money in Middle Eastern Countries' September.

Dehbod, A. (1963), 'Land Ownership and Use in Iran', CENTO Symposium on Rural Development, Tehran. Denman, D.R. (1973), *The King's Vista.*

Geographical, London.
Diehl, L. (1982) *Small Holder Farming Systems with Yam in the Southern Guinea Savannah of Nigeria.* Deutchegesselschaft fur technishe Zusammengarbeit, 126, Eschborn.
Dornbusch, R. (1980), *Open Economy Macroeconomics.* Basic Books, New York.
D'Silva, B. and M. Rafique Raza (1980), 'Integrated Rural Development in Nigeria', *Food Policy,* November.
Essang, S.M. (1977), 'Impact of Oil Production on Nigerian Agricultural Policy', *Indian Journal of Agricultural Economics, 22,* 2.
Evans, H.D. (1986), *A Political Economy of Trade and Development.* Wheatsheaf Press, Brighton.
Fadama Survey Report (1984), Zaria.
FAO (1974) *Production Yearbook.*
FAO (1982a), 'Farm Support Prices for Wheat', Committee on Commodity Problems, Rome.
FAO (1982b), *Fertilizer Yearbook.*
FAO (1983a), *Technical Co-operation Programme, Water Resources, Irrrigation and Reclamation Development Policies, A Food Producltion Strategy–Nigeria.*
FAO (1983b), *Statistics on Agricultural Support Price, 1972–81.* Rome.
FAO (1985), 'Irrigation Sub-sector in Nigeria: Progress, Status, and Future Outlooks', paper presented to FAO's Irrigation in Africa Seminar, June.
Federal Office of Statistics (various years), *Rural Economic Survey.*
Forrest, T. (1982), 'Recent Developments in Nigerian Industrialisation', in M. Fransman ed. *Industry and Accumulation in Africa,* London.
Gorgani, Mansur (1971), *The Economy of Gorgan, Gonbad, and Dasht.* Safialishah, Tehran.
Graham, R. (1979), *Iran – The Illusion of Power.* Croom Helm, London.
Hays, H. and J. McCoy (1978), 'Food Grain Marketing in Northern Nigeria: Spatial and Temporal Performance', *Journal of Development Studies,* January.
Hooglund, E.J. (1982), *Land and Revolution in Iran, 1960–1980,* University of Texas Press.
Hooglund, E.J. (1973), 'The Khwushneshin Population of Iran', *Iranian Studies,* Autumn, p. 230.
IBRD (1972), *Report on Iranian Agricultural Sector.*
Idachabe, T. et. al. (1981), *Rural Infrastucture in Nigeria, Basic Needs of the Rural Majority, Vol. 1 Main Report.* Federal Department of Rural Development, Lagos.
Ike, D.K. (1977), 'Estimating Agricultural Production Functions for Some Farm Families in Nigeria', mimeo.
IMF (1983), *International Financial Statistics.*
IMF (1984), *International Financial Statistics.*

Industrial and Mining Development Bank of Iran (IMDBI) (1970–8), *Annual Reports*.
International Fund for Agricultural Developments (IFAD) (1985), *Nigeria – A pre-identification desk study*.
Jakubiak, M. and M. Dajani (1976). 'Oil Income and Financial Policies in Iran and Saudi Arabia', *Finance and Development*, December.
Jazayeri, A. (1986), 'Prices and Output in Oil-Based Economies: The Dutch-Disease in Iran and Nigeria', *IDS Bulletin*, November.
Johnston, B. and P. Kilby (1975), *Agriculture and Structural Transformation*. Oxford University Press, New York.
Jones R.W. (1965), 'The Structure of Simple General Equilibrium Models, *Journal of Political Economy, 73*, December.
Khosravi, K. (1976), *Rural Sociology of Iran*. Payman, Tehran (in Persian).
Kilby, P. (1981), 'A Review of Prospects and Problems in Agro-Allied Industries in Nigeria', World Bank, mimeo.
Killick, T. (1983), *IMF and Stabilization in Developing Countries*. ODI, London.
King, R. (1981), 'Co-operative policy and Village Development in Northern Nigeria' *Rural Development in Tropical Africa*. Macmillan, London.
Kirk-Greene, A. and D. Rimmer (1981), *Nigerial Since 1970*. Hodder and Stoughton, London.
Kravis, I.B. *et al.* (1981), 'New Insights into the Structure of the World Economy', *Review of Income and Wealth*, December.
Kravis, I.B. *et. al.* (1982), *World Product and Income*, UN Statistical Office, Johns Hopkins, University Press, Baltimore.
Lagemann, J. (1977), *Traditional African Farming Systems in Eastern Nigeria*. Weltforum Verlag, Munich.
Lambton, A.K.S. (1969), *The Persian Land Reform, 1962–66*, Clarendon Press, Oxford.
Leamer and Stern (1970) *Quantitative International Economics*. Boston.
Lewis, Stephen R., Jr. (1982), 'Development Problems of the Mineral Rich Countries', research memorandum Series, no. 74, The Centre for Development Economics, Williams College, Massachusetts, October.
Lipton, M. (1977), *Why Poor People Stay Poor*. Temple Smith, London.
Mackinnon, R.M. (1976), 'International Transfers and Non-Traded Commodities: The Adjustment Problem', in D. Leipzigen, ed., *The International Monetary System and the Developing Nations*. AID, Washington DC.
Madjd, M.G. (1983) 'Land Reform and Agricultural Policy in Iran', Cornell/International Agricultural Economics Study 16.
Mayer, W., (1974), 'Short Run and Long Run Equilibrium in a Small Open Economy', *Journal of Political Economy, 82*, 5.
Ministry of Co-operation (1976), *A Report on the Activities of the Ministry in 1975*. Tehran (in Persian).

Mitra, A.K. (1972), *Production, Production Requirements, Costs, and Returns of Crops,* Ibadan.
Mo'meny, B. (1980), *The Agrarian Question and Class Struggle in Iran.* Payvand, Tehran (in Persian).
Morgan, D. (1979), 'Fiscal Policy in Oil Exporting Counties, 1972–78, IMF Staff Papers, March.
NEDECO (1975), Kano River Project, Kano Project Area, Part IV, Agronomy.
National Planning Office (1981), *The Fourth National Development Plan, 1981–85.* Lagos.
Ngozi, Okongo-Iweala (1982), *Developing Financial Institutions in Nigeria's Rural Areas: Some Farm-household Perspectives.* World Bank.
Norman, D., D. Pryor and D. Gibbs (1979), *Technical Change and the Small Farmer in Hausaland, Northern Nigeria.* African Rural Economy Paper, 21, Dept. of Economics, Michigan State University.
Norman, D. et al. (1982), Farming Systems in the Nigerian Savannah, Westview Press, Boulder, CO.
Nowshirvani, V.F. (1978), *Production and Use of Agricultural Machinery and Implements in Iran: Implications for Employment and Technical Change,* Plan and Budget Organisation, Tehran.
Nurkse, R. (1964), *Problems of Capital Formation in Underdeveloped Countries.* Oxford University Press, London.
Okafor, F.C. (1975), 'Labour Shortage in Nigerian Agriculture', *NISER,* University of Ibadan.
Okigbo, P.N. (1981), *Nigeria's Financial System.* Longman, Harlow.
Osuntogun, A., ed. (1983), *Rural Banking in Nigeria.* Nigerian Institute of Bankers, Lagos.
Oyejide, T.A. (1975), *Tariff Policy and Industrialization in Nigeria,* Ibadan University Press, Ibadan.
Palmer-Jones, R. (1984), *'Irrigation Projects in Nigeria',* Irrigation Problems in Tropical Africa, Oxford Centre for African Studies.
Pick's Currency Yearbook (1977).
Pick's Currency Yearbook (1979).
Plan and Budget Organisation (1975), *Iran's Fifth Development Plan 1973–78 Revised: A Summary.* Tehran.
Price, O.T.W. (1975), *Towards a Comprehensive Iranian Agricultural Policy,* World Bank, Agricultural and Rural Development Advisory Mission, September.
Razavi, H. and Vakil, F. (1984), *The Political Environment of Economic Planning in Iran, 1971-1983.* Westview Press, Boulder, CO.
Rimmer, D. (1984), *The Economies of West Africa.* Weidenfeld and Nicholson, London.
Sadigh, P. (1975), 'Impact of Government Policies on the Structure and Growth of Iranian Industry', Ph.D. thesis, Dept., of Economics, University

of London.

Safi-nezhad, J. (1977), *Buneh,* 3rd edn. Toos Publications, Tehran.

Salmanzadeh, C. (1981), *Agricultural Change and Rural Society in Southern Iran.* Middle East and North African Studies Ltd., Canterbury.

Salter W. (1959), 'Internal and External Balance: The Role of Price and Expenditure Effects', *Economic Record, 35,* 226–38.

Sano, H.D. (1983), *The Political Economy of Food in Nigeria, 1960–1982.* The Scandinavian Institute of African Studies, Uppsala.

Schatz, S. (1981), 'Nigeria's Petro-Political Fluctuations', *Issues, 11 – 2.*

Schatz, S. (1977), *Nigerian Capitalism.* University of California Press, Berkley.

Shafaeddin, S.M. (1980), 'A Critique of Development Policies Based on Oil Revenues in Recent Years in Iran', Ph.D. thesis, University of Oxford.

Snape, Richard (1977), 'Effects of Mineral Developments on the Economy, *Australian Journal of Agricultural Economics,* December.

Statistical Yearbook of Iran (various years), Tehran.

Stock, R. (1977), 'The Impact of the Decline of the Hadejia River Floods in Hadejia Emirate' in J. Van Apeldoorn ed., *The Aftermath of the 1972–74 Drought in Nigeria.* Centre for Social and Economic Research, ABU, Zaira.

Stopler, W. and P.A. Samuelson (1947) 'Protection and Real Wages', *Review of Economic Studies 9.*

Swan, T. (1960), 'Economic Control in a Dependent Economy', *Economic Record, 36* 51–66.

Taylor, L. (1982), 'An Economic Model of an Oil Exporting Country', MIT mimeo.

Udo, R.K., (1975), *Migrant Tenant Farmers of Nigeria.* African Universities Press, Ibadan.

Udo, R.K. (1983), *Food Production and Agricultural Development Strategies in Nigeria.* Hitotsubashi University.

Udo, R.K. (1984), *A Comprehensive Geography of West Africa,* Heinemann Education Books, Ibadan.

USDA (1981), *Nigeria: Agricultural and Trade Policies.*

Van Wijnbergen, S. (1984), 'Inflation, Employment, and the Dutch-Disease in Oil Exporting Countries: A Short-Disequilibrium Analysis', *Quarterly Journal of Economics, 99,* 2 May.

Wallace, T. (1981), 'The Challenge of Food: Nigeria's Approach to Agriculture, 1975–80', *Canadian Journal of African Studies, 2,* 5.

WARDA (1981), *Rice Production and Marketing in Nigeria,* Occasional Paper no. 3, January.

Watts, M. (1983), *The Silent Voice,* University of California Press, Berkley.

William, G. (1978), 'State and Society in Nigeria', *Afrografika,* p.164.

Williams, G. (1983), 'Why Is There No Agrarian Capitalism in Nigeria', paper presented in IDS Seminar, University of Sussex, Summer.

Wilson, R. (1931) *Capital Imports and Terms of Trade,* University Press,

Melbourne.
World Bank (1974), *Nigeria–Agriculture Sector Memorandum.*
World Bank (1975) *Land Reform.* Sector Policy Paper.
World Bank (1979) *Nigeria–Agriculture Sector Review.*
World Bank (1981a), *Sokoto ADP, Appraisal Report.*
World Bank (1981b), *Accelerated Development in Sub-Saharaan Africa: An Agenda for Action* (known as the Berg Report).
World Bank (1981c), *Report on Non-Oil Exports of Nigeria.*
World Bank (1983), *World Development Report.*
World Bank (1984a), Nigeria–Agricultural Sector Memorandum.
World Bank (1984b), *ADP Issues Paper,* April.
World Bank (1984c), *World Development Report.*
World Bank (1984d), *Nigeria–Agricultural Pricing Policy,* Report No. 4945-UNI, October.

Index

Abalu, G.O.I. 141
Abuja 113, 115
Afshar, H. 78
Agbola, A. 150, 151
Agricultural Development Bank 82
Agriculture, Iranian 73-4, 86-7
  see also Land-ownership prior to reform; Land reform; Prices, costs and production
Agriculture, Nigerian 123-4
  cocoa 151-2
  groundnut 148-9
  impact of agricultural policies 135-42
  irrigated cash crops 124
  maize 144, 146-7
  marketing 134, 135
  millet and sorghum 147, 152
  National Accelerated Food Production Programme (NFPP) 140-1
  Operation Feed the Nation (OFN) 141
  Palm Produce Board 149, 150
  see also Agricultural Development Programmes; Credit constraint; Farming systems; Green Revolution Programme; Irrigation; Labour constraints; Land tenure system; Palm oil and palm kernels; Prices, production and income; Rice
Agricultural Development Programmes (ADPs, Nigeria) 135, 138, 139-41, 146
  adoption by 'progressive' farmers 140-1
  Funtua ADP 140, 141
  Gombe ADP 140
  objectives 140
  Sokoto state ADP 141
  success 141
  World Bank loans 140
Alizadeh, P. 70
Amuzergar, J. 78
Ashraf, A. and Banuazizi, A. 76, 78, 83
Avromovic, D. 61
Awa, N.E. 132, 134
Awojobi, A. 115, 136

Carter, R.C. *et al.* 137
Caves, R. and Jones, R. 31n, 32n
Centre for Agricultural Marketing

Development (CAMD) 92
Chavanian, M.M.H. 85
Clough, P. 144
Collier, P. 144, 147
Credit constraint (Nigeria, agriculture) 132-4, 135

Dehbod, A. 75
Denman, Mr 78
Development strategy and composition of expenditure (Iran) 39-48
   budget 42, 43
   cost of living, rise of 45-6
   deflationary programme 47
   economic factor, 'scarce' 42
   Fifth Development Plan 40, 42, 43, 45, 85
   interventions, crisis mode 47
   investment by sector, fixed 40
   merchandise imports, total 44
   military expenditure 41
   oil revenue decline 46
   port bottlenecks 45
   price fixing 46
   resources, political misallocation of 43-4
   spending spree 42-3, 44, 46
   subsidies 41
Diehl, L. 129, 144, 147
Dornbusch, R. 4
D'Silva, B. and Raza, M.R. 141

Economics and Social Statistics Bulletin 115
Essang, S.M. 1

Fallah, S.V. 90
Farming systems (Nigeria) 127
   cassava-growing 129, 152
   cultivation of lowland flood plains (*fadama*) 129, 138, 139, 140
   intermediate system 127, 129,
   northern grain-based 127, 129
   southern root crop-based 127, 129
   yam-growing 127, 129, 144, 152
Fedayeen Organisation 77, 81, 83, 84
Food and Agriculture Organisation (FAO) 89, 91, 136, 138, 139, 141, 144
Trade Yearbook 94
Forrest, T. 119

Gahreman, Mr 98
GDP and components (Nigeria) 116-21
Gorgani, M. 82
Government expenditure, pattern of (Nigeria) 111-15
Government interventions 156
   consumer subsidies 30-1
   content 21
   credit policy 27-8, 70
   crisis mode, within 20, 47
   fiscal and monetary policy 21-5
   form 20-1
   land-ownership policy 27
   planning mode, within 20, 47
   pricing policy for food crops 28-9
   sectoral allocation of investment 25-7
   tariff policy and import control 29-30, 70
Graham, R. 41, 45, 64
Green Revolution programme 135, 141-2
   agro-service centres 142
   Credit Guarantee Scheme 142
   National Council for Green Revolution 142

Hays, H. and McCoy, J. 134
H.N. Agro-industry of Iran and America 84
Hooglund, E.J. 76, 78, 81

Idachabe, T. *et al.* 126
Ike, D.K. 125
International Agribusiness Corporation of Iran 85
International Fund for Agricultural Development 140
International Monetary Fund (IMF) 102, 107, 117
Iran
   1970s 35-6
   Central Bank 39, 42, 92
   'Companies for Utilisation of Land

Downstream of Dams' law 84
comparisons with Nigeria 105, 120-1
Industrial and Mining Development Bank of Iran (IMDBI) 64, 66, 68, 70, 72, 97, 101: loans 66, 67, 68, 96
life expectancy 105
Ministry of Agriculture 83, 92
Ministry of Co-operation 78, 83
Ministry of Finance 42
Ministry of Industry 68
Ministry of Rural Affairs 77
Pahlavi foundation
per capita income 105
Plan and Budget Organisation (PBO) 40, 41, 42: Co-ordination and Supervision Division 43; Planning Division 40-1, 42
population 105
Shah 39-40, 41, 42, 46, 47, 68
Statistical Centre 74, 92
Statistical Survey 78
Statistical Yearbook of Iran 50, 63, 91, 102
Iran–California Agribusiness Company 84
Irrigation, large-scale (Nigeria) 135, 136-9
Backolori Dam Project 127, 136, 137
compensation for dispossessed 137
cropping pattern, conflict over 137-8
Fadama Survey Report 139
failure 138
Kano I River Project 136, 137, 138
River Basin Development Authorities (RBDAs) 135-9 *passim*, 142, 145
shadoof, use of 139
Sokoto Irrigation Project, 127, 136, 138
South Chad Irrigation Project 136
Tiga Dam 127, 138
waterlogging 137

Jakubiak, M. and Dajani, M. 38, 39
Johnston, B. and Kilby, P. 1, 87

Keyhan International 46, 63

Khosravi, K. 80, 82
Kilby, P. 149
Killick, T. 102
King, R. 133
Kirk-Greene, A. and Rinner, D. 107, 119
Kravis, I.B. *et al.* 157

Labour constraint (Nigeria) 30-2, 135
comparison between north and south 131
effect of universal primary education 131
index of wages and prices of major export crops 130
shortage of labour 131
wages 130-1, 132
Lagemann, J. 129
Lambton, A.K.S. 77
land-ownership (Iran) 75-6
boneh (water co-operative) 76, 77
crown lands 75
family tenancies 75
ghanaat (water channels) 76, 77, 80-1
owner-cultivators 75, 76
private holdings 75
public domain 75
religious or public purposes, land for 75
sharecropping 75, 76, 80
land reform (Iran) 74-5, 76-81
agribusiness 84-5, 86
boneh, disappearance of 75, 80
farm corporations 82-4, 85, 86
first phase 76, 77-8
land excepted from 76, 77
land grants below subsistence sufficiency 79, 86
landless workers 80, 86
land market, distortion of 81, 86
nasaq, distribution by 77
numbers affected 77, 78
property rights, uncertainty over 75
rural co-operatives 78, 81-2
'rural development poles' 85
second phase 78, 79
sharecroppers, favours 80

structure 78-9, 86
third phase 78, 79
village-collective becomes family farming 73-4
land tenure system (Nigeria) 124-7
ambiguity re rights to land 125
'backyard' gardeners 126
gayauna (family head) system 125
land distribution 125
land-use decree 126-7
large capitalist farmers 126
migrant-tenant farmers 126
peasant resistance 127
small farms 124-5, 126, 127
Lewis, Arthur 86
Lewis, S.R. Jr 1
Lipton, M. 87

Madjd, M.G. 78, 79, 85, 87, 88, 89, 91, 93
Manufacturing, Iran 57-8
brick production 68, 69
carpet production 88
cement production 68
cigarette production and imports 97
credit, role of 70
decline in 62-3, 65, 71
effect of post-1973 boom 58, 62-71
electricity, shortage of 63-4
food processing 67
footwear 66, 69
government influence 59
growth in contribution to GDP 58-9
import decontrol 63
industrialisation 58-62
industries, fast-growing 69
national economic profitability rating 61
non-metal mining products 68-9
price controls 64-5, 60
production, types of 59
restrictions, lifting of foreign exchange 63, 64
share-divestiture scheme 6, 64
strikes, 63
tariff protection 59-60
textiles 65-6, 119
transport equipment and metal products 67-8, 69
wage increases 63
Mitchell, Duncan 85
Mitra A.K. 151, 152
Mo'meny, B. 78, 83

NEDECO 147, 148
Ngozi, Okongo-Iweala 132, 134
Nigeria
Central Bank of Nigeria 111, 113, 133, 134
comparisons with Iran 105-6, 120-1
Federal Office of Statistics (FOS) 116, 118, 143, 144
Fourth National Plan 116, 130, 132-3, 136, 138, 140
life expectancy 105
map of 128
Ministry of Water Resources 136
National Planning Office (NPO) 112, 114, 115, 116, 131
per capita income 105
population 105
Technical Committee on Producer Prices 134
Third National Plan 112, 136, 138
see also GDP and components; Government expenditure; Oil revenues
Nigerian Gross Domestic Production and Allied Macro-Aggregates 117
Non-oil GDP and components (Iran) 47-8
GDP, changes in sectoral distribution of 53-5
GDP, non-oil 48, 54-5
construction, wages and prices in 53
growth of non-oil GDP, sectoral 51-3
labour, price of 50-1
price increase of main sectors, cumulative percentage 49
prices, changes in relative 48-50
Norman, D. *et al.* 131
Nowshirvani, V.F. 76
Nurkse, R. 42

Obasanjo, Gen. 113
Oil revenues (Iran) 36-9

capital flows abroad 37
currency fluctuations 37-8
dependence on oil exports 36
Oil revenues (Nigeria) 107-11
    deficit, domestic 109
    expenditure, growth in 109-11
    foreign earnings, rise in 107-8
    imports, increase in 108
    money supply 109-10
    naira, appreciation of 111
    tariffs, relaxation of 107-9
Okafor, F.C. 131
Osuntogun, A. 132, 133

Palmer-Jones, R. 138
Palm oil and palm kernels (Nigeria) 149-51, 152
    exports 149-50, 151
    labour requirement 151
    wild palmeries 149
Prices, costs and production after 1973 (Iran) 87-9
    barley 93
    cotton 95-6
    oilseds 96
    potatoes, onions and tomatoes 98
    pulses 97-8
    rice 91-3
    Rice Impact Programme (RIP) 92, 93
    sugar-beet 93-5
    sugar imports 94
    team 96
    tobacco 97
    wheat 89-91
Prices, production and income (Nigeria) 142-3
    export crops 148-52
    food crops 142-52: *see also* separate crops
    production figures 143

Razavi, H. and Vakil, F. 40, 41, 43, 46
Ricardo, David 158
Rice (Nigeria) 144-6
    imports 144-5
    prices 144-5
    production, cost of 145
    production, rise in 144
    swamp rice (*fadama*) 145
    upland (rain-fed) rice 145
Rimmer, Douglas 150, 151
Rural Economic Survey 125

Sadigh, P. 63
Salmanzadeh, C. 83, 84, 85
Salter, W. 4
Sano, H.D. 126, 133, 139, 140, 141
Schatz, S. 111, 112, 113
Shafaeddin, S.M. 41
Shagari, President (Nigeria) 113
Shel-Cotts Agribusiness of Iran 84
Sunday Times 112
Swan, T. 4

Theoretical framework 1-2, 155-9
    adjustment for internal and external balance 9-10
    core adjustment model 4-10
    demand conditions 6
    disequilibrium 8-9
    Dutch-Disease model 47, 67, 71, 73-4, 75, 95, 101, 139, 150 155, 156
    factor incomes adjustment 11-12
    factor markets and output supply 4-6
    internal and external balance 7-8
    money, relative prices 12-16
    non-traded goods 2-4
    role of capital inputs and cost of production 16-19
    sources 1
    see also Government interventions

Udo, R.K. 126, 142
US Consumer Price Index 46
US Department of Agriculture (ASDA) 118, 143-4, 146, 147

Wallace, T. 138
Watts, M. 142
West Africa 113
West Africa Rice Development Association (WARDA) 144, 145
Williams G. 137, 138, 144
Wilson, Mr 4
World Bank 60, 78, 105, 110, 129, 131, 133, 135-9 *passim*, 141, 143

APMEPU Baseline Survey (Nigeria) 130
loans for Nigerian agriculture 130